## Links

*www.bobfairbrotherauthor.com*
*https://x.com/bobfairbro*
*www.instagram.com/bob_writer_fairbrother/*
*www.facebook.com/BobFairbrotherTheAuthor/*

# Dedication

To Janet for countless years of support.

# 1

# Holroyd

The first November rains of winter were washing away an arid year with a vengeance, causing the gutters to run red with Saharan sands.

Miss Holroyd arrived at her school desk early, taking a tissue to her splattered shoes.

She'd planned quiet time to get ahead of the day before the hordes of pupils descended; only Harry Butler is at her desk with his irritatingly non-symmetrical football badges on his jumper. Until now, his only contribution to her classes had been an air of indifference.

And she thought she had an idea why he is here. A friendly colleague, Jan Sanderson, had told her to *watch out* for Harry, and that a student teacher had left her post in a hurry, because of Butler's *attentions*. Sanderson must have sensed her veiled euphemisms were annoying her because she quickly added that the student teacher hadn't revealed what had happened, possibly fearing claims of sexual harassment would damage *her* prospects.

'It's 2046 but how times have regressed,' Holroyd had replied.

Butler leaned onto her desk, looming over her.

His stomach protruded over his belt. A corner of a white shirttail arrowed towards his right thigh from under his sandy-coloured jumper, slovenly yet fashionable.

Miss Holroyd forced a professional smile. 'Harry, what can I do for you?'

Butler moved around the desk, a further advance into her personal space. She caught invasive notes of peppermint gum, blended with the sharp tang of body odour.

'I really like you, Miss.' He reached a hand towards her, destination inappropriate.

Droplets of dirty water fell from him, catching the weak morning light.

'You are obviously the type of male I would be interested in, Harry,' said Miss Holroyd, each word buttressed with steel, contradicting the persona she usually projected.

The boy preened. Then frowned. 'Wait a minute, you're being *sarcastic!*' The word crowded his mouth, as if he wasn't used to saying it.

'How perceptive of you, Harry.'

His face scrunched in bafflement at this unexpected resistance, hand now frozen in pre-grope, assurance flailing like a sail on a windless sea. Harry's face flashed red with anger, fists clenching.

She rose sharply from her chair, forcing Butler to step back, an unbalanced retreat from the space he had so confidently occupied.

'And if I am not interested? Then what, Harry?'

'Eh?'

'I suspect this consideration has never occurred to you before. Maybe it's a different sort of lesson you require.'

The boy's eyes widened; an insight gained. 'You're

*threatening* me! I'm reporting you, you... you're a teacher... you can't say that!'

'And your behaviour is acceptable?' she hissed her question.

Harry Butler turned, exiting the classroom in a hurry. 'You're raving mad, Miss.'

His Pavlovian response required him to add a 'miss', even after his casual abuse. *That's education in action.* She thought as she swiftly followed him.

*Harry may be a victim of circumstances, and without access to rehabilitation, he'll probably trample a path through many lives, aspiring to emulate his father, the paedophile. They both deserve to be on her list.*

She followed Butler out into the hallway and called out to him.

'Harry, stop, I've not finished with you yet.'

He halted at the top of the staircase, then made an exaggerated effort to turn to face her, shrugging contemptuously.

'I ain't worried about you, miss. You're just a temporary teacher. You can't do anything, can ya? None of you can. And you... I'm gonna catch you outside, then we'll see how smart you are.'

The boy stood close to the edge of the stairs, careless, cocky.

'My dad told me...'

'Your *dad* told you to do many things, didn't he? Like how to bring those boys to your home.' Her blunt words hit him like ice-cold sponges.

Harry Butler looked shaken at having a secret part of his world spoken out loud. He stoically stuck to his mantra '... th..that I'm to tell him if you lot take any liberties, he'll...he'll be down here so fast you lot won't know what fuckin' hit...'

'He's dead. This morning. In a road incident.'

Blood vacated Butler's face, now a graven image of shock.

'What? You're... you're lying!'

'I know I could've broken this news more softly to you. But I feel we have a rapport, Harry.' She stepped closer to him. 'And *rapport* means we understand each other.'

'I... I don't believe you.' Defiant, tearful, burning eyes, hoping.

'I assume you are referring to your father's death and not about our *rapport.* He'd been wearing a blue Chelsea scarf with badges like yours, only marked with his blood.'

The boy's eyes stared into the past, perhaps recalling what his dad had been wearing that morning. Harry's shoulders slumped, the fight leaving him like foul air from a punctured tyre, his tears breaking ranks.

'No, it's not true,' said Harry, guard down, hit by the first punches of grief.

Then Holroyd gave him a firm shove.

Harry Butler snatched reflexively for the staircase railing, only grasping stale corridor air. His mouth open, eyes shocked wide, as he travelled backwards. Gravity took the baton, pressing down on him.

Holroyd watched Butler tumble in a series of inelegant somersaults. He lay in a heap at the bottom, body misshapen. His scream echoes around the wide, vacant staircase and halls. *The noise would bring others.*

Holroyd hurried down the stairs, crouched to check on the boy, and found him still breathing. *If he got immediate medical attention, he might yet live.*

'I can't have Master Butler surviving and reporting all kinds of stories, can I?' she whispered conspiratorially to him, her breath on his ear, her blonde hair draped across his cheek with a proximity that he had probably dreamed of.

His eyes conveyed a desperate fear. She thought his spinal nerves were probably damaged, speech beyond him.

Miss Holroyd had scant seconds before anyone arrived but said, 'I don't enjoy killing teenagers, Harry, but when they are budding little monsters, I think I can make an exception.'

Harry's terrified eyes are full of the dreadful knowledge of imminent death. Pleading.

'Don't worry, Harry, it's a rhetorical question, which means no answer is required.'

She pinched his nose and placed a hand across his mouth. She felt the vacuum on her hand as he vainly sucked for air, his chest convulsing. A few rolls of his eyes and the sixteen-year-old Harry Butler was gone.

## 2

# DCI MacGillivray

DCI MacGillivray stood up from his office chair, back complaining. He stretched and stared out through the cracked, dirty police station window. The rainy, flat grey morning light gave him a surly greeting.

A stray shaft of sunlight broke through the cloud cover, hurting MacGillivray's eyes and making his bones ache. He'd always preferred winter when the landscape had been softened by the day's dying light, when he could *breathe* in the dark.

Now, the first rains at least offered relief from the heat from another challenging year of dry, cracked earth. Of trees that caught fire. Of soil blown away to dirty brown dust from super-farmed fields. Now, there is a gritty constant in the air, joined by the desert sands, whipped up by mistral winds that turn the sky a shell pink.

Now, the rains signalled the start of extreme storms and flooding.

He'd lacked the momentum to go home last night to an apartment of unpacked removal boxes. Instead, he'd tried to turn his attention to the neglected police cases on his desk: piles of printed paper - a sign of a world without computers. And yet he could not distract himself, could not escape his mind's compulsion to rake over Yvette and Lucia's deaths. And of the killer, Sam Marsh, who, six months ago, had walked free from court. His celebrating entourage of family and legal representation had accompanied him, the latter wearing their smug victory with a tinge of embarrassment. As if acknowledging their complicit part in releasing the guilty back into the wild.

MacGillivray also had his shame for what he'd done, causing a stress line of pain to cut across his stomach.

Before his wife and daughter's deaths, MacGillivray had strived to deliver justice in the right way. But then he'd got *that* call of a double homicide, when he'd heard his address.

When everything changed.

He cast his mind back. He didn't remember driving; just arriving there, finding a line of police cars blocking his usual parking spot. Another jolt of reality that nothing was as it should be.

DI Crozier met him in his driveway. When she put a hand on his shoulder - a simple act - it had taken away any vestige of hope; *it was his family dead in there.* His shoulders strained against dropping. His island of shock had created a strange, watery distance from the real.

'You can spare yourself this,' Crozier had said.

'I have to see.'

He hadn't needed his key to his door because it was open. The traffic of Scene of Crime Officers, all offering sympathetic glances. *It's strange how death gave strangers license to roam in your private world.*

He'd crossed the threshold of this place he'd never call home again.

His Yvette and Lucia lay in the hallway. Faces pale,

white, leaden. Lacerations of knife cuts had opened channels in their flesh, freeing blood to flow among the floor tiles of green and adding their own type 'O' red diamonds.

His wife's body, an arrow pointed towards their daughter, hands stretched, trying to protect.

His sweet daughter. The home-made bracelet, delicate, at home on her wrist. He remembered it banging against his wrist as he held her hand as she skipped. A simple joy, never to be repeated.

Her eyes stared at nothing, death giving her a cold indifference–that was never like her.

*Could his heart freeze? He thought so now.*

That's when his good philosophy for delivering justice slid.

Crozier said gently, 'CCTV footage from the street corner has placed Sam Marsh and his son Troy in a car approaching the scene at about the time of your... the victims' deaths. For MacGillivray, it was too much of a coincidence.

*He knew what he had to do.*

A THUMPING ENGINE outside drew MacGillivray back to the present.

He followed the sound to the window; a tractor chugged past, churning accumulated mud, belching black diesel fumes, trying its best to befoul the wet air.

MacGillivray stepped back, seeing his reflection in the semi-dark glass under the flickering fluorescent light. Flecks of grey in the fringes of his brown, unkempt hair, his stubbled face, he looked older than his thirty-eight years. He wore a tired-looking shirt scuffed around the neck. His suit carried a dull sheen, fibres crushed. *Not a brilliant look,* he conceded.

He disliked self-pity, his stubborn pride holding himself together, but sometimes, he couldn't see beyond this bitter, black dog despair. *What's he going to do without his wife and child? How would he live with his failure? His shame?*

His office door jabbed open. Detective Inspector Crozier walked in without ceremony, whistling *Summertime.*

She took off her dark blue police peak cap hat, revealing her jet-black hair tucked into a bun, like black brush strokes against her coffee and cream complexion.

Crozier's whistle cut off mid-tune, eyeing MacGillivray's crumpled clothes. 'Didn't you make it home last night?' She gave an experimental sniff, her nose wrinkling.

*Yes, I could use a shower,* he thought, embarrassed, then replied. 'Eh, no.'

He rubbed his eyes to massage clarity in and exhaustion out.

'So,' she said. 'You have a lot to feel bad about. Losing your family. And the killer, Marsh, getting off. Poor you, right?'

Her words were a firehose of iced water smashing into MacGillivray.

He looked at her, eyes wide, startled. Her style is not unfamiliar territory, but even for DI Lerato Crozier, her words are brutal. He feels his blood pulse. Now unable to breathe, like the room's oxygen being sucked out, anger igniting.

*It's my bloody pain. I'm entitled to it. Leave me be.* He ached to shout at her. He fought the urge to lash out, to send the case files flying.

The feeling surprised him; it was so untypical of him. He slammed his hands down on the desk, grappling to control himself. Despite his anger - because of his anger - his words dammed in his throat.

Crozier gave a merest nod of satisfaction. *Perhaps he had given her something she wanted to see: a spark of fight, of life, in him, perhaps.*

And she's not finished.

'Look around you. Other people have lost loved ones. While the economy is crashing and the wild swing of the climate gets worse, they live in fear in a dog-eat-dog world as the criminal elements run riot. Forgive me for trampling over your sensibilities, but it's our job to protect them. So, could I respectfully ask you to get your head out of your backside and help? Sir.'

*He knew what she's doing: it's a peculiar brand of kindness, well camouflaged.*

He could feel a shadow of a smile on his face, his jaw muscles twitching as they relaxed.

'Your bedside manner could do with some work,' he croaked, pulling breath in. 'Has it slipped your attention that I am your boss?'

'I can only protect you for so long. Coffee?' she offered.

He nodded. Crozier left the room, her tall figure set in her default determined pose, her act of care in the community temporarily suspended.

*Crozier was right; she had been covering his backside. But even she couldn't work miracles forever. Her ambition was to climb higher, yet she stayed loyal to him, refusing to throw him to the dogs.*

He sat back in his chair and opened Marsh's well-thumbed file. He massaged the web of flesh between his thumb and forefinger.

*She's right; he must focus on other cases and pull his weight.*

*But if he found evidence, he'd still try to bring Marsh down.*

A knock at the door broke into his thoughts.

'Come in.'

PC Grey came into MacGillivray's office. Stale cigarette air wafted in with him. Grey, one of the many who smoked. It had become the cool retro drug, if you could afford it. Not that MacGillivray's an anti-smoking zealot; it's just that the smell made him feel nauseous, and he had a selfish desire to avoid getting cancer.

MacGillivray was yet to fathom Grey out. He came highly recommended; the government assigned emergency help, parachuted in to help clear police station backlogs. He gave off the air of a team player, but something about him which did not fit. Crozier had told MacGillivray that Grey's hair reminded her of *wet black slates on a stormy day*–she'd been making fun of him–but had she inadvertently *got* him?

'There's been a report of a pupil's death at the Forêt School. A fall downstairs, sir,' said Grey as Crozier walked in.

'Thank you, we're on our way,' said MacGillivray.

'There is something else you should know. The victim's father was the fatality in the car wreck this morning,' said Grey.

'The paedo?' said Crozier.

'Unproven, DI Crozier,' said MacGillivray wearily.

'Sorry, you're right. The dead, unproven, paedo.'

Grey's eyebrows raised.

'Given time, you may get used to DI Crozier's *unique perspective*,' said MacGillivray.

'Dark humour is a way of surviving the wave of human tragedy we face, sir,' said Grey.

'I absolutely agree,' said Crozier, eyes sparkling. 'Of course, it is awful that these things happen. But if we allow everything to get to us, we might not function. We'd risk going crazy,' she revealed a hint of brilliant white teeth as she spoke.

'This is you *not going crazy*, right? Let's go, shall we?' said MacGillivray, shaking his head.

---

MACGILLIVRAY COAXED THE AGEING, dark blue BMW i3s towards the Forêt school. Rain hammers on the roof. A feeble waft of warm air drifts from the vents inside the cold interior.

'Are you sure the heater's working?' said Crozier, hunching her shoulders, rubbing her hands vigorously together. *It isn't so cold, though it felt so compared to the higher temperatures of only last week.*

'It's only a quick trip, you'll survive that, won't you?'

'Are you implying that women feel the cold more than men? The patriarchal wedge of superiority driven just a little further, perhaps?'

'No, of course not... I meant...'

'I know what you *meant*,' she said, smiling. He felt himself unhooked and dropped back into the water.

'Anyway, you are a few years younger than me. It's me who should feel the cold.'

'Yes, opa.'

Crozier mentioned to MacGillivray that she'd never lived in the Netherlands but had picked up some of the language from her Amsterdammer mother. They were her memorial words for keeping her mother's memory close.

And MacGillivray had even absorbed a few Dutch words. '*Opa?* I am not quite old enough to be your grandfather, thank you.'

He turned on the car radio to catch the news. He missed the local radio station, but with independent broadcasting gone, the BBC feed is the only game in town.

The radio presenter spoke, an inflexion of north-eastern brogue embroidering her words.

'The Chancellor of the Exchequer is today delighted to announce that the UK inflation rate has plateaued and could soon head downwards.'

'It is still eighteen per cent, but any good news is welcome,' said Crozier.

The voice on the radio continued, 'Meanwhile, as riots escalate in the cities and the growing crime epidemic presents a sterner challenge to the police, the government has released good news: The National Security Surveillance and Action (NSSA), responsible for investigating police corruption, has increased its mandate to deal with syndicate crime and will also manage ad hoc gatherings of excessive numbers.'

'Manage?' said Crozier. 'That's code for pulling anyone off the streets, even those that voice a peaceful opinion. The government wants the illusion that they're in control of the country. And just before the general election, too. It's all about staying in power.'

MacGillivray had some sympathy for a government dealing with a desperate situation, *but is the NSSA the way to change it?* Rumours followed the organisation, and they were not all good, with reports of a heavy-handed approach growing by the day. And the organisation's encouragement of people to report their neighbours' *suspicious behaviour* with the promise of units of electricity or free food seems tasteless. The government couches it as self-policing, but it feels more like an attempt to split society and a cynical attempt to take the focus away from the government. Not that the public often gathered in Hawham, the town's despair brought its own brand of apathy.

'The government has announced the release of an AI machine to help make policing safer for all concerned. They've been fast-track testing a prototype that'll provide basic curfew policing duties in the capital.'

'Wow,' said Crozier. 'Isn't the Government's budget too tight to even maintain normal policing, much less provide funds to develop this line of enforcement?'

'Perhaps they've interest from external investors, from big business. A sign of lobbyers finding new, creative ways of getting their way, of protecting what they have and hold,' said MacGillivray.

'Steady, DCI MacGillivray, that sounds practically rebellious for you.'

They passed a council driverless two-seater car, abandoned on the street after the network collapsed. Traffic had compacted accumulated summer dust and dry earth around it, where rainwater now carried the mush away to the gutters.

The voice from the radio continued, 'Experts say the malicious strains of malware that signal blocked Wi-Fi, disrupted the internet and damaged GPS satellite systems will take more time to be corrected.'

‘They’ve been saying that for almost a year. So, we stay in the technological dark age for a while longer,’ said Crozier.

MacGillivray could see the pitch through the gate of Hawham Football Club’s stadium: a pool of water now covering its entirety. A sign, white with red letters, read Hawham Football Club: Season Suspended Until Further Notice. *One more addiction that has been closed off to the public.*

Smoke trails rose, issuing from chimneys, opposing the vertical downpour.

Crozier said, ‘So many people are forced to hibernate in their homes, deciding between food and heating. As if it’s the government’s passive way of culling the burden of sick, old, and vulnerable in the country.’

‘That is cynical. Crozier.’ Though he, too, had his doubts. The collection of bodies in the town morgue had quietly grown, many of those souls unmissed, now gone and forgotten.

The radio presenter said, ‘During the rainy season, the government advises the public only to venture out if necessary and keep a close watch on the vulnerable.’

‘The BBC can’t give weather forecasts until they have contact with satellites again, so they issue vaguely general warnings of rain or heat, instead,’ said Crozier.

The Forêt school came into view; yellow police ticker tape surrounded the entrance, like inappropriate bunting, fluttering.

MacGillivray parked in a corner of the car park next to other police transport. The area resembled a beach against the lapping water blossoming from a failed drain. He switched off the engine, and the noise of rain battering the car roof became more pronounced. Even in this torrential downpour, he caught the familiar hint of rotting waste in rubbish piled outside of homes nearby, waiting for the striking dust persons to return to work.

After enduring a year's heat, it's odd to see their breaths making fine white plumes in the cooler air. They walked under the protective wing of a concrete canopy, passing temporary prefab classrooms, stagnating into permanency when the extension wasn't built.

'It's got the look of a detention centre,' said Crozier.

Her comment is too close to the truth for MacGillivray's comfort. The school, one of seven secondary schools in Hawham, had morphed into a dumping ground for those with educational and behavioural problems - a depressing production line of the police's future workload.

The entrance front is an expansive foyer, part of the stalled aspiration of becoming an academy. A solemn silence of a near-empty school met them; the floor smeared with weather-trailed feet of hundreds of children and the smell of boiled cabbage and antiseptic, transporting him back to his school days–*some things stay constant.*

MacGillivray and Crozier bypassed the scanning machine, flashing their ID cards at the security staff.

'It's hard to believe there used to be a time when scanning for weapons wasn't required,' said Crozier.

PC Jacobs greeted them. 'The SOCOs are still at the Butler crash scene, but they'll be here soon.'

'Thank you, Jacobs. Let's look at the site, anyway.' Jacobs is short; his trousers are too small, he is hanging shy of his feet, and his pants are too big to fit his waist. His black hair needs cutting, and he carries a crown of pale white flesh on the top of his head. His face has a sallow complexion, and spider veins anointed the side of his nose; an excessive red flush implies a man who was never far away from his next drink.

Crozier, with disposable plastic gloved hands, lifted the cover to see the victim. A youthful body, now lifeless–a terrible contradiction. Some fresh scuff marks on the wall from Harry Butler's fall were barely discernible from day-to-day wear. Nothing screamed incompetence or malicious intent.

They returned to find a sore-eyed receptionist hovering, waiting for their return. She confirmed the headmaster would meet them shortly. A grandiose coat of arms hung on the wall, quartered with a prince, a tree, a shield, and a sword. Headmaster Fawngate appeared from his office, like a wary lord, protective of his fiefdom.

'Headmaster Fawngate. I am sorry to meet you in such circumstances. A French Prince gifted a forest,' said Fawngate, nodding towards the object of MacGillivray's attention. 'It's since been cut down for school and residential housing. I often wish they'd kept the forest.'

Crozier flashed MacGillivray an arch eye as they were ushered into Fawngate's office. 'A tragic accident,' Headmaster Fawngate said, as if remembering the grim business of Harry Butler's passing. He placed himself behind his desk, more comfortable in his seat of power.

'We'll need to complete our enquiries before we can ascertain if it's an *accident*, sir,' said MacGillivray.

'Of course,' said the Head, bristling at the correction.

But MacGillivray thought the Head was right; there's nothing to suggest foul play, even if the boy's father had died that morning.

'What sort of lad was Harry Butler?' MacGillivray asked.

The headmaster pursed his lips and steepled his fingers.

'He wasn't without his challenges. A slow, uninterested pupil, who made up for his lack of academic industry with his close attention to his fellow pupils, shall we say?'

'A bully.'

'Indeed.'

'Is there any truth in the rumour he assaulted a student teacher?' asked Crozier.

Fawngate's eyes flared. 'An unsubstantiated claim, with no action taken. I would be grateful if you could be discreet when discussing this matter; we've our reputation to protect.'

*Said without a trace of irony,* thought MacGillivray.

The receptionist knocked and ferried a tray of tea and biscuits on the Head's invite.

MacGillivray sipped his builder's tea and studied the stiffly sitting Fawngate.

'When did you learn of Harry Butler's fall?' asked Crozier.

'We heard Harry's scream. Mrs Holtershaw, my receptionist, and I came out of our office to see Miss Holroyd and the boy lying at the bottom of the stairs. I checked for a pulse, but he was dead. Mrs Holtershaw called the services. I don't envy your job of breaking the news to the parents. Awful business,' said Fawngate.

'Unfortunately, his father died this morning in a car incident.'

'Oh,' said Fawngate, his empathetic vocabulary exhausted, then electing to say indignantly, 'Why wasn't I informed earlier?'

'The body didn't have any ID on it. And, without technology, getting access to car records takes time.'

'Oh. It's a shambles without the internet, isn't it?' said Fawngate.

MacGillivray bypassed the comment, saying, 'That'll do for now, Mr Fawngate. We want to see Miss Holroyd,' said MacGillivray.

'I agreed for her to go home early. On full pay, of course,' he said with a magnanimous tilt of the head. 'However, I think she is still here.'

'With the school being shut anyway, that is still benevolent of you, sir,' said Crozier.

MacGillivray flashed a warning at Crozier to reign in her facetiousness - though, he couldn't deny, her ability to ruffle people's feathers got results.

Fawngate rifled through the papers on his filing cabinet, nettled that his part in this tableau of misery had been so easily dismantled.

'Here's her file,' he snapped, passing the brown folder to MacGillivray, ignoring Crozier.

It didn't help Fawngate's mood when MacGillivray commandeered his office for the interview.

---

IT'S Miss Holroyd's green eyes that MacGillivray noticed first. They carried a mortician's professionalism.

According to the file, she's in her early thirties yet could pass for younger, lines absent from her pale white skin. She had an air of calmness, at odds with someone who had just encountered the trauma of finding a dead child.

'Please take a seat, Miss Holroyd.'

She lithely moved across the room. Sat boldly in the middle chair of the circle, fronting Holroyd and MacGillivray.

'How are you after your ordeal?'

'I fared better than the boy.'

'Yes, indeed. You were first on the scene of Harry Butler's death?'

'I was.'

'Did you see him fall?'

'I did. I came out of room 2A. He'd been running and tripped down the stairs. I got to him, but he was gone. Mrs Holtershaw and Fawngate arrived. They called for the services.'

An uncluttered, unemotional appraisal of a child's death if ever MacGillivray had heard one. 'Fawngate's given me time off to get over the shock. You can contact me at home if you need to ask further questions.'

MacGillivray glanced at Crozier, her right brow upturned. *Who's in charge here?*

'Thank you, Miss Holroyd. We've no more questions for now. If you decide to leave town, please contact us first.'

She gave a curt nod, then left.

Crozier said, 'That is one cold morsel. Would you want her as a teacher?'

'Perhaps she is what they need in this bear pit of a school.'

'Or people to dispense compassion and understanding?'

'I didn't expect you to be peddling those, Crozier.'

She gave him a quizzical look, then uttered, 'Grappig,' the Dutch word for *funny*, emphasised with a guttural *g*.

'Should we check out Butler Senior's Road Traffic Incident just in case there is a connection?' said Crozier.

'That is a good idea,' said MacGillivray. *Visiting the RTI might be a dead end. But maybe Crozier had a point; they should take a look.*

# 3

# Ash

Miss Holroyd had been at his last school at Warburn-on-Sea, Ash was certain. *But she'd given no sign she'd remembered him. He'd not been in her classes there, so why should she? But wouldn't it be friendly to at least say hello to her?*

He came in early to catch her on her own, to avoid embarrassment in front of everyone if she didn't remember him. Only, as he approached her room, he saw, through the glass panel of the door, Harry Butler talking to their teacher. *What was Butler doing here so early? And is he* coming on *to Miss Holroyd? It looked like it,* he thought. He leaned towards the door to hear better.

He'd never heard a teacher talk to a pupil like she did; it was funny how she shot him down.

A red-faced Butler was marching straight towards him.

Ash dashed into another classroom. He'd not been at the school long but knew Harry Butler had a bad temper. It wouldn't be good for Ash's health if Butler knew Ash had seen him crash and burn.

Harry walked past, followed closely by Miss Holroyd, who was telling him to stop.

Ash had peeked around the corner, trying to hear what they were saying.

Then he heard Harry's scream.

HE COULDN'T BELIEVE IT. *Had he just seen Miss Holroyd push Harry to his death?*

Shaken, frozen in situ. He agonised over what he should do.

Mr Fawngate's voice came from the staircase. He didn't feel he could ever say anything to *him*. Ash made his decision and took the other staircase, avoiding everyone.

LATER, he stood outside waiting for his friend, Steve.

*Is it a coincidence that a boy had died at his last school, too? And hadn't Miss Holroyd been the one to find him first there as well?*

*Is Miss Holroyd a serial killer?*

*That's bloody crazy.*

His thoughts were big enough to distract him, to forget to keep out of Tommy Marsh's radar. Now Ash found himself hemmed in under the bus stop by the bully Tommy and with the bonus of that wanker Eddie Bridges with him.

'Hey, climate refugee,' shouted Tommy Marsh, unphased by the rain.

*Climate refugee. Shouldn't that describe people getting away from danger?* But Tommy used it as an insult, a putdown. As if it's Ash's fault the sea had flooded his home year after year with ever rising tides. The summers were spent drying their home out, fixing damage that his dad said no insurance company would now cover. The almost broke council wouldn't build the defences to hold back the sea because, his mum said, *'Why would they throw good money after bad?'* — whatever that meant. A sandbag-fortified shallow wall built by the village had been effective against the gentler summer tides, but the winter seas picked holes in their defences. Salty water continuing to find its way into their homes.

In the end, his mum and dad gave up and moved out. Leaving behind their home and friends. To move into their shitty mouldy little terraced house. In this shitty little town. And -him at this shitty little school. It hadn't helped his parents' mood when a neighbour told them that the river Haw could burst its banks any time. *Can't the water just stay away?*

'For a bargain five credits a week, I can offer you protection,' said Tommy Marsh.

*Protection from Tommy, that is.*

Tommy blathered on as if Ash had no choice but to take it.

*It didn't matter; he didn't have money to give him.* So, he did the one thing Tommy least expected, that Ash least expected: he delivered a couple of quick, hard punches to Tommy's jaw.

*It's a dangerous move. But it's worth it for that moment, to see the shock on Tommy's face.*

Tommy's sidekick, Eddie, is stunned. But he rallies and punches Ash on the side of his head, causing stars to appear behind Ash's eyes. Ash retaliates, swinging his elbow, feeling the give of Bridges' eye. The boy doubles over, holding his face.

Marsh recovered from Ash's punches, smashing his right fist into Ash's face, knocking him to the ground.

An Anglo Bubbly Bubble gum wrapper and dog shit are at Ash's eye level. *The rain is actually bouncing.*

He hadn't noticed rain do that before, but then again, he'd had never seen it at ground level before. He tried to return to his feet. His mouth tastes of copper, blood pooling around his teeth. Tommy urges Bridges to 'kill 'im.'

Ash gathered his courage for the fight.

# 4

# DCI MacGillivray

'What are you doing?' demanded MacGillivray, arriving with Crozier.

The three boys froze. Staring at the new arrivals.

'What's it to you?' MacGillivray recognised the speaker as Tommy Marsh, the son of Sam Marsh.

'Two against one - is that fair?'

Tommy, short but already powerful, closed the distance between them, saying, 'My old man's gonna hear about this.'

Then he brightened with malicious glee, recognising MacGillivray.

'I know you. How is your family? Oh, I'm sorry, you don't *have* one, do you, copper?'

Blood surged in MacGillivray's head. The world reduced to one impulse - *to rip this little bastard's head off.*

He heard his name spoken by Crozier, enough to relock the constraints of duty.

'I suggest you go,' said Crozier to the boys.

Tommy withdrew, puffed up with his impact on MacGillivray, stabbing a finger towards Ash Heath. 'I'll see you later.'

MacGillivray felt sick at his loss of control. He

needed to concentrate on something tangible—their target, the boy.

Strawberry hair hung over the lad's collar. *He can barely have a kilo to spare on him, his troubled eyes searching for an escape.*

'Are you alright, son?'

Crozier produced a tissue for Ash to dab his bloodied mouth.

'Yes, thanks,' he said, watching Tommy and his sidekick walk away.

'What was that about?'

'It's nothing,' he said, his eyes communicating alarm with their brightness.

'It usually isn't,' replied Crozier.

The boy shrugged.

'What's your name?'

He shifted his feet.

'Don't worry; you're not in trouble.'

'Ash Heath.' The words dealt out reluctantly.

'I'm DCI MacGillivray and this is DI Crozier. Don't get the wrong side of the Marshs.'

'Sure.'

*It's probably too late for that tip. Ash's tight swallow suggested he knew, too.*

But MacGillivray recognises *obdurate,* and he is looking at it now. He had a feeling that he'd not seen the last of Ash Heath.

## 5

# Ash

The police officer, MacGillivray, nodded at Ash as he drove away in his old blue beamer. MacGillivray seemed alright to Ash, for a police officer, but he couldn't tell him what he might've seen, *because who trusts the police?*

Ash is aware of the thumping in his chest, of his heart galloping after his fight. Now, there was time to think, to be sick with the dread of trouble.

The school is nearly deserted except for one lone figure who had appeared at the entrance to the school. A heavy hooded coat, trousers, and hat made it difficult to tell the gender. *Was it a girl, maybe? Going by the way she moved. Yes, she's a girl. And she is moving towards him.*

'Hello, I am Angelica.' *She speaks and occupies space so confidently, in a way he thought he never would.*

Her hair poured out from under her Red Riding Hood; only she reminded him more of the wolf.

*It's rude to stare. But he did it anyway. By the look of her clothes, she came from a nicer part of town. What did she want here?*

'Hi, I'm Ash.' His mouth felt like it had Novocain injected into it.

She seemed amused. Her laugh echoed, filling his head, to be played on repeat later.

'Are you lost?' Ash wanted to keep her near for a little longer.

Angelica unzipped her coat. She revealed her neck, framed by a crisp shirt. Her loosened tie denoting her membership of the private school, an alien from another world. He detected a scent of lemons and summer. She reached into her coat pocket and produced a light blue slip of paper, offering it to him, still warm. Ash read her name and number, elegant handwriting, unrushed. A drop of rain dropped from his nose. It spread out on the paper where it had landed, catching the blue ink, spreading like a dark invader.

'No, not lost. But you can help me,' she said. 'By calling me later. Today,' she emphasised, stepping within a foot of him. Citrus hung around him, only deeper now.

Something stirred to life in his stomach. He'd never felt this before.

The girl put her arms around his neck, pulling his mouth onto hers, kissing him hard. Warm, crushed cherries pushed against his lips. He felt weightless, floating like the bubbles in the cheap, fizzy wine his parents allowed him at Christmas.

'Why me?' said Ash after she stepped back from him, part of him annoyed at himself for asking such a lame question.

'I saw you stand your ground with Tommy Marsh. No one does that. And I must confess I noticed you before already.' Her smile showed a hint of perfect teeth. Her eyes, electric blue, eclipsing everything around her. Then Angelica turned and walked away.

Before, there'd been greyness. Now, excitement

and possibility exploded into his world. He looked down at the paper, thrilled, and yet he felt caught.

'WHO'S THAT?' Steve's voice brought him back unwillingly from his spell.

Ash looked up at his friend.

'Angelica Richards,' said Ash, reading the slip of precious paper.

'What was she after?'

'I'm not sure.'

'Hmmm, she seemed to know,' Steve said, grinning, holding out for a fist bump.

'Shut up, you idiot,' Ash said, unable to keep a straight face, feeling the strain in his jaws relax.

They ambled to the school's covered inner yard, its plastic roof melodic with the sound of rain. Ash took a size three football out of his bag, let it fall to the ground, and struck it with his instep. The tattered white ball hit the wall satisfyingly, bouncing neatly back to him.

Angelica and the fight had made him forget about Harry Butler, but seeing Steve brought that back to him. He wanted to tell Steve about Harry, but the more time passed, the more he doubted what he'd seen. *It felt easier just to forget it.*

'What happened?'

'What?' shocking Ash into thinking he's talking about Harry.

Steve pointed to the marks on Ash's face.

'Oh, yeah. Tommy Marsh and Eddie Bridges.'

'Oh.' He nodded. 'They're twats. Will your folks say anything?'

Ash thumped the ball against the brick wall,

causing splinters of brick mortar to desiccate and fly. 'No, they'll be too busy arguing.'

He magnanimously teed up the ball for Steve, who attempted a half-volley. The ball skewed off his foot, ingloriously settling in the corner of the yard's gutter.

*Anyway, telling his parents would break a code. It's a strange honour, when he thought about it, being bullied, beaten up, then not saying anything about it: what am I honouring?*

'Do you know where Tommy lives?' Ash asked.

'Yeah, why?'

'Can you show me?'

'You're playing with fire.'

'Look, I am frightened, okay? But I'm in for endless hassle if I don't do something. It's fine if you don't want to be involved.'

'I'm in. But what are you gonna *do*?'

'If I can scare Tommy enough, with no one knowing I scared him, he might leave me alone.'

'That's all? Oh, that's all right then. You're nuts.'

'I need to catch him when he is out on his own.'

'Yep, it's official, you're nuts.'

'THAT'S THEIR PLACE, big, isn't it?' said Steve.

'Yeah,' *scarily large*. 'If they have this sort of money, why go to a crap school like the Forét?'

'Maybe he doesn't want his kids to go to a posh school because they'll show him up?' said Steve.

Ash nodded, thinking it believable that some parents would be happy to hold their kids back.

They couldn't go through the gate and risk being seen. Ash picked a spot next to the Sussex stone wall under the sparse cover of an oak tree. He gave quick, wary turns of the head. Looking for a threat in this well-off suburbia. 'Give me a leg up.' asked Ash.

Steve, muttering darkly about nutters and madness, created a stirrup with both hands. He propelled Ash up onto the wall.

Feeling the flint wall's sharp edges against his hands, Ash hitched himself over the wall. Breaths coming jerkily from the effort. He jumped down, cursing silently every breaking twig underfoot as he rushed towards the garage. Its door stood a foot's width ajar. Ash risked a peek inside.

TOMMY MARSH SWUNG punches at a punch bag tethered from a beam across the ceiling.

'I can't believe I let the bastard get away with it.' Ash heard Tommy say, 'I'm gonna get him back tomorrow. He's gonna friggin' GET it. I can't believe he fought back!'

*He's talking about me.*

Another volley of punches struck the heavy bag, making minor indents, interspersed with grunts.

'What are you on about?'

Ash jolted back at the sound of the new voice, then peered again to see Tommy's older brother Sid enter from the inner door.

'That new kid made a fool of you,' said Sid.

Tommy, irritated, hammered more punches into the bag.

'He's lucky. We were gonna get into him...'

'And the police stepped in. Yeah, I know,' said Sid, interrupting him.

Ash could see Tommy's face reddening.

'Remember, you're carrying the family name, our

honour, our reputation. Dad don't want people getting cocky, believing they can take us, okay?'

'Yeah, Eddie will help me.'

'Make an example of Ash Heath. Remember, fresh blood is coming along, and they'll want to take what is ours. So, deliver on the promise of pain, and you'll keep them fearing you.'

'*Promise of pain*?' said Tommy, trying to sound bored, but really hiding his interest. But even he could hear the belligerence in his voice.

'You can carry a reputation only so far, then they must be reminded that pain is the answer if they don't keep in line.'

*It's nasty,* but it sounded smart to Ash. He understood better now why the Marshs had a hold, why people didn't mess with them.

'Yeah,' said Tommy sullenly. 'Are you finished? Then clear off and talk to your friends on your Ham radio. Do they know who you are?'

*Ham radio–what's that?*

'Family reputation is all that matters,' said Sid, ignoring Tommy's poke at him.

'I've already rung Eddie; he's going to meet you at the river bridge,' said Sid to a furious-looking Tommy. 'Because we don't want our business being overheard on the phone, do we?'

*Is he talking about phone tapping? Do people do that? In little Hawham?*

Tommy stomped out of the garage into the now light rain. Ash could hardly believe his luck–here's his chance.

Ash waited for Sid to return to the central part of the house. Then followed Tommy through the entrance gate, waving to Steve to follow.

'This is crazy,' Steve whispered when he caught up with Ash.

Ash didn't need reminding. He just hoped Tommy didn't glance back.

Half a mile later, Tommy turned left, following the river that registered a new high against the tide posts. He stopped and lit a cigarette by a bridge. 'He's meeting Eddie here,' said Ash.

'Great.' said Steve, without enthusiasm, as they hid behind a nearby van.

*When will I get a better time to catch Tommy alone?* Ash thought. *Eddie could turn up, but it's a risk he had to take.* Ash took off, heading for his quarry. He couldn't just turn up and face Tommy. He had to have something to give him an edge, to frighten him; that was why he'd found an old cricket bat in his dad's garage, its weight hardly outweighing his fear.

Marsh swung his arms and stamped his boots, trying to keep warm under a large tree. The smoke from Tommy's newly lit cigarette caught on the chilled wind, casting a trace of nicotine for Ash to smell, and Marsh's fledgling smoker's cough is muffled by the trees.

A wind whipped plastic bags into tight whirlwinds. One caught on Ash's legs, almost tripping him. The road darkened with the cover of pregnant clouds, devoid of artificial street lighting, helping to mask his approach.

Then, out of the gloom, Ash saw a tallish man dressed in black, heading towards Tommy.

'Damn,' he murmured, turning back to rejoin Steve.

'Who's that?' said Steve.

'Dunno.'

'I don't like it,' said Steve, chewing at his index fingernail.

Tommy drew on his cigarette, exhaling smoke in the man's direction.

'What do you want?' Ash heard Tommy snarl.

'You, Tommy,' said the stranger.

'Why don't you clear off?' But despite the aggression, Ash sensed uncertainty hiding behind Tommy's aggression.

The man pulled a wallet from his pocket and flashed it at the boy.

'I'm Tommy *Marsh*.' Exaggerating his surname, like a spell to ward off evil.

The man stepped closer.

'We've got the filth in our pocket, so why don't you fuck off?'

*Police? Is the man a police officer?*

The man took another step forward. Disturbingly calm.

Tommy, sensing an animal threat from the man, shrank. His shoulders fell away, head leaning back. *Tommy is frightened.*

Ash looked around for help. The traffic lights on the road were red. But there's no traffic to stop, and the street is empty. *What* could *they do? Why didn't Tommy run for it? It's as if he's stuck.*

Steve gasped as the man grabbed Tommy's arm, fingers like a bear trap, steel teeth snapping shut.

*Tommy had been the shark in their waters. But now there were bigger predators to fear in the world.*

The man placed a cloth over his nose and mouth. Only now, Tommy tries to strike out. *Too late*, thought Ash.

The man's too strong, leading him to the bank of

the river. The boy's struggles lessoning. *Had he been drugged?*

The man removed the cloth.

Ash glimpsed Tommy's face. He was sobbing, rain sluicing his snot trails. The man let go of Tommy and he fell towards the water. *It's strange how Marsh didn't use his arms to stop his fall.* The slap-splash sound of his body hitting cold black water echoed under the bridge.

Ash wanted Tommy hurt. He'd even wished him dead. *But he'd not meant it, not this.* It didn't take long for the bubbles to stop rising from Tommy's face under the water. Tommy's body floated, snagging on a marooned tree.

Ash spotted a second body, also caught in the tree's web of limbs. It's a grey-faced Eddie Bridges, his face looking blankly up to the sky, floating loosely.

Ash glanced at Steve, who was wide-eyed and unbreathing. Ash realised he was holding his breath, too. He felt his brain fusing yet holding a fascination at what he saw. And a disgust at himself for having that fascination.

The man looked around, his gaze a dark beam from an evil lighthouse, causing them to duck. Ash knew that if the man saw them, he would kill them.

The killer walked along the shallow bank to the road. He wasn't even rushing. The rain fell harder, an unwitting accessory helping to cover signs of murder.

After a few stunned moments, Ash noticed rain-water leeching into the knees of his trousers and felt water trickling down his shins.

'Shouldn't we do something? Tell someone?' Ash said, his heart beating in his ears.

'Are you crazy? That's the *police.* We can't trust anyone. Let's get out of here.'

‘Wait,’ said Ash, ‘look. Someone’s walking their dog.’

# 6

# DCI MacGillivray

PC Hanbury greeted MacGillivray and Crozier at the river crime scene.

Crozier once described Hanbury as handsome, *with his Roman nose and peaches and cream complexion.* MacGillivray could see that, and he thought him a bright lad, too, honest, willing, hardworking and with a rare sense of empathy. But if the police station closed, Hanbury could face the scrap heap in this desperate economic climate at twenty-six years of age, which would be a waste.

'Their bodies were spotted floating by a dog walker who called it in,' said Hanbury, nodding towards the open body bags, ready for them to inspect.

'It's Tommy Marsh and Eddie Bridges–the two that were bullying the lad at the school, right?' said Crozier.

'Yes.' *Marsh's son.* Empathy for losing a child clashes with a sense of karma that the universe had served out to Sam Marsh. MacGillivray knew he had to be better than this.

'We'd better break the news to their parents,' said MacGillivray, aware of Crozier's x-ray gaze on him.

He reached between his thumb and forefinger, massaging intensely.

MARSH LIVED in a better part of town. A detached house, large garage, fences, trees, and copious lawns demarcating the space between them and the neighbours.

'Are you okay doing this?' said Crozier.

'We're thin on the ground for staff, so the rules get stretched. Let's get on with it.'

Crozier looked unconvinced.

The glossy red front door opened, revealing an antique grandfather clock sitting against the wall in the hall. It looked out of place amongst the modernity of the house's décor. The pendulum swung, reminding MacGillivray of humanity's march to the grave.

MacGillivray recognised one of the Marsh boys, Sid, who answered the door.

The teenager eyed MacGillivray, the database of his brain fitting the person in front of him to a memory he recognised.

'What do you want, *copper*?' His inflexion makes it sound like a vile swear word.

MacGillivray sucked in a painfully patient breath. 'Are your parents in?'

Sam Marsh arrived behind Sid. Sid's father appeared to have lost his neck as if the muscles had risen in revolt: *Sid's evolutionary path foretold - and they already resembled two thuggish Russian dolls,*

'Who is it, Sid?'

'The filth. The one who thinks you killed his family. And he looks like shit. But the other one, she don't look half bad, to be fair.'

‘I’ve told you before about disrespecting women, Sidney.’ Marsh said, setting eyes on Crozier, ‘But my son’s got a point. I wouldn’t want AI replacing you if you believe all the stories out there.’

He looked back to MacGillivray, his dismissal of Crozier, ironic. ‘Now, what can I help you with? Is it something I should be worried about?’

*Apart from me hunting you down?*

‘Regrettably, it is, sir,’ MacGillivray said as measuredly as he could to the killer of his family.

‘It might be a good idea to speak to your wife, too,’ Crozier said evenly, ignoring the misogynistic baiting, causing Marsh's eyes to narrow.

‘What’s this about?’ demanded Sid, also sensing something unusual, replicating a snake flicking its tongue to taste the air, finding something not to its liking.

‘Yes, what *is* this about?’ said Marsh, his sharp voice betraying his wariness. Nothing faux about his concern, now.

‘It’s about your son, Tommy.’

MARSH LOOKED ASHEN FACED, rendered much older. His wife’s face, though, set to chiselled stone. Despite his shock, Marsh put an arm around her shoulders. It’s an awkward, stilted effort, which made it kinder—something MacGillivray did not want to see.

Crozier approached Sadie Marsh.

‘I am allergic to the filth, so sling your hook,’ she spat, eyes flashing hatred over the thin line of her mouth. Crozier retreated.

Sid stood up from sitting on the arm of the sofa as

if the act would make it easier to take in the news. 'Little Tommy? Dead? That's not true,' he said.

The front door opened, then slammed shut.

MacGillivray recognised Troy Marsh, the oldest of the three sons. Twenty-five, tall, lean muscles. *His long blond hair, smart clothes, and manicured nails try to hide the thug, but makeovers don't easily mask menace.*

MacGillivray recalled an incident at the Carpenter's Arms in town. He'd reviewed footage from the pub. A young lad had joked that Troy looked 'disco ready.' Troy laughed along. He even struck a dance pose to amuse everyone in the bar. But MacGillivray recognised the relaxing in Troy Marsh's body, seeing the internal coil bunching, eyes focused.

Troy smashed the joker in the face, picking up the first thing that came to hand, a heavy metal ice bucket. The victim parted company with many of his teeth, blood staining the pub's busy Axminster carpet. Then Troy had leapt on the unfortunate victim, sending a volley of punches into his face. Luckily, his friends pulled him away before he pummelled the man to death.

*The joker refused to press charges, saying 'it'd been a misunderstanding'.*

*No doubt Sam Marsh had squared it away with money. Marsh; always saving his family. If only he could've done that,* MacGillivray thought of himself bitterly.

*And Troy, a rottweiler, on his father's lead, with the lead showing signs of fraying.*

'What's going on?' said Troy Marsh, picking up the vibe.

'Tommy's dead,' Sam Marsh said, jaw twitching, nodding towards MacGillivray and Crozier. 'They think Tommy and Eddie were messing about on stones in the river.'

Sid looked his older brother in the eye and said, 'I don't believe that for a minute. I don't care what those bastards think–it's a job that been done on our brother.'

Did MacGillivray see a glimmer of respect for Sid in Troy's face? And in Sid, behind that tough front, he saw a hunger for more of that respect.

'Who did it? The Leighs? But we've sliced up our business interests between us.'

'Perhaps times are hard enough for them to risk a fight with you to get a bigger share of those *business interests,*' said Crozier.

'Don't say anymore,' said Sam Marsh, flashing a warning to his sons.

'Did you do it?' said Sadie Marsh, jumping up, baring teeth at MacGillivray. He noticed her crooked, stained molars–*she might be tough, but she's evidently wary of the dentist.*

'Love, this'll get us nowhere,' said her husband, placing his placating hand on her arm, which she shrugged off like a fly.

'Sod off, Sam. My little boy is *dead,*' she snarled, her eyes somewhere between incandescent and mournful for her loss.

'Let's go,' said MacGillivray to Crozier. *What had he been thinking coming here?*

'Yeah, get out,' said Sadie Marsh.

'You'll need to identify your son,' said Crozier, directing the question at Sam Marsh.

'I'll be there in an hour,' said Marsh as he slammed the door.

## 7

# Ash

Ash didn't know what to do. *Should he report what he knew? But what if the killer* is *the police? It's bad enough with bullies at school, and now there are serial killers, dodgy police and criminal gangs to worry about, too.*

Ash needed self-protection *but the cricket bat wasn't going to do. He'd had some when he owned a knife. He'd saved his credits to buy one from a trade-in shop. The black-handled knife with its serrated blade in the window had looked sexy.*

*He'd hidden the blade in a box in his cupboard at home in his bedroom under a pile of old games, which his parents had found. He'd been angry at them for snooping around his things, breaking his trust.*

But they said knives *attract trouble* and that they give you a *reputation.*

*He got that, and he didn't really want a knife. But he needed a reputation — to keep the thugs away.*

*Then he had an idea.* 'Come on,' he said to Steve.

'Where to?' he replied.

'My home to collect a few things.'

THE CATAPULT LEANED against the bedroom wall, a simple weapon made of metal and rubber, along with a bag of marbles. *At least it would be something. Ash's grandfather had given it to him–perhaps that's why his father had not taken it away, too–because of sentimental value.*

*His old man had given strict instructions not to use it against another human being. Well, things change. He'd have to deal with that shit storm later.*

He delved into his bottom draw–*it's still there!* The redundant tech of his old phone, along with its charger. *Not long ago, everyone had a phone, could contact anyone almost anywhere, and had so many apps on the piece of tech. And people used it to take selfies.*

*And with this phone camera, it would be easier to prove what he found.*

'Let's go to the café; I don't want to answer awkward questions when my parents return.'

ALTHOUGH IT WAS late on a Friday, most of the café tables were full. They ordered, sat down and sipped their hot chocolate. Ash missed the small flake that used to come with the drink, which is no longer available because of shortages, but the cocoa and sugar hit is fantastic. He'd read somewhere that sugar is good for shock–*and didn't he have some?*

Ash said, 'I know we shouldn't tell anyone, but we've seen *two murders.*'

Steve shook his head quickly, 'My parents won't do anything even if I told them.'

Ash couldn't imagine telling his anything, either. *Nothing important, anyway.*

*They wouldn't want to do anything with what he'd see; they'd just want to stay invisible, too.*

'I know we can't trust the police, either, but I think I could tell that policeman with the Scottish name, he seemed alright.'

'Not a great idea. What if it *was* the police who'd killed them? And anyway, it's grassing!'

'Grassing on someone who is killing people.'

Steve looked down at his drink, preoccupied with adding more sugar.

'Bang goes your theory about Miss Holroyd being a serial killer,' said Steve.

'Yes, you're right, it was a man that killed Tommy and Eddie.'

A lorry trundled slowly past on the high street, pushing waves of encroaching water against the kerb stones, causing little water jets to punch into the air. They reminded Ash of the seas hitting the sea wall in his old town. Vibrations from the lorry rattled the café tables, making cups shake on their saucers. Peering through the drizzle, Ash noticed a man on the other side of the street, then he jumped up, electrified.

'It's him!' he said.

'Who?' said Steve, following Ash's line of sight.

'Tommy and Eddie's killer. That's him, look.'

The man wore a cap, a heavy-duty black raincoat down to his knees, boots, and black trousers. Ash had only seen him in poor light by the river, yet it was sure it was him. He looked like death walking.

'You're bloody right, it is him.'

'Come on,' said Ash. He opened the door to the café, allowing the cold to rush in, to mumbled complaints from the café's customers, and allowing the smell of black pudding and rock cakes to escape outside.

They tentatively jogged along the rain-glazed pavement that reflected light from shop windows, keeping the killer in view. And when the man glanced back, they dropped out of sight.

They tracked the killer to a block of flats on the east side of town. The man took the porch entrance, lights going on behind the curtains on the first floor. Ash nibbled at his lower lip, looked around, and they both slipped into the main entrance. Taking the stairs, Ash saw the giveaway signs of fresh wet patches leading to number twelve.

*They know where the killer lives.*

'What now?' whispered Steve when he got back.

Ash put his finger to his lips, waved Steve towards the stairs, and said, 'Let's wait over there.'

Ash led the sprint to get out of the rain under the bus shelter's cover.

It had been almost an hour. Ash felt numb from the insidious, damp cold. Having spent so long in the heat and humidity all year, the coolness of the rainy season hit hard.

'I am so bloody cold,' said Steve, his teeth starting a little tap dance. 'How long are we gonna hang around here?'

'Stamp your feet and get your blood moving. At least the rain is easing off,' said Ash, as rivulets of water, carrying detritus down the gutter, slowed a little.

'Hey, look,' said Ash, 'He's leaving the building.'

Ash's relief that their wait was over was replaced by the anxiety of doing something dangerous, the pulse in his neck throbbing harder. He pulled his old mobile out of his rucksack.

'Why have you got that old piece of junk?'

'It's got a camera on it.'

Steve's eyes widened. 'Genius!'

Ash snapped off a couple of shots, catching the

killer as he turned to face them. Terror and relief passed through Ash at not being seen.

The high streets tapered off, merging into suburbia before melding into the industrial estate by the river.

Keeping the man just in sight, Ash saw a boy further ahead of the killer. The sight of the boy seemed to prompt the man to quicken his pace.

'I know him, it's James Leigh,' said Steve, uttering his name ominously, 'Another family to avoid.'

The killer pulled a white cloth from his coat pocket.

*Could what happened to Tommy and Eddie be happening here?* Ash thought. Guilt still gnawed at Ash about Tommy's death. He told himself that if he had tried to help, he would've been killed, too. Though, deep down, he knew fear had frozen him.

He wanted to be like the heroes in his granddad's old collection of Avengers comics. *Did they feel terrified and sick at the first sign of danger?*

*It had to be different this time. But what to do? Perhaps he could distract the killer.*

He pulled the slingshot from his ruck, his 'Y' shaped piece of metal with an elastic band tied to two prongs.

'A catapult and an old phone? What else have you got in there?'

'Marbles.' Ash undid the drawstrings on the white cloth marble bag and pulled a red-eyed glass sphere, slotting it into the rubber pouch. He tugged, looking down the stretch rubber line, straining for the sweet spot, aiming at the security box window across the road. He released the band, feeling it snap back, drizzle droplets flicking off. Ash and Steve both held their breaths as the marble took flight. The window gamely tried to accommodate the red-flecked orb, but the glass could no longer defy the laws of physics, exploding into a myriad of imperfect pieces, spasming into the office.

Ash felt a thrill, a release of tension, as the noise of destruction burst from the impact. At that moment, he knew he should run away–but he had to see what happened next.

The killer snapped towards the jagged noise; he pulled a handgun from his pocket–*he's got a gun?* Ash could see a security guard staring out the window–*had he hurt him?*–The gunman fired quickly, hitting the guard.

'Oh no,' said Ash, disbelief blunting his senses.

'Oh shit,' said Steve.

'Run,' Ash shouted, urging the rooted, open-mouth James to do so, too.

The gunman turned towards a now-running James, pulling the trigger again.

James jerked, falling off the river path and disappearing into a bank of deep mud.

'He...he shot him!' said Steve.

*Now, the gunman's chasing the escaping, staggering security guard.*

Ash could see blood on the guard's coat, even from this distance. On the ramp leading up to the warehouse, the guard reached the door, fumbling with his keys.

'Hurry,' Ash shouted.

The guard had found the right key, inserting it into the lock.

*He's going to make it.*

*The gunman stopped, took aim and fired.*

Ash heard the guard's grunt of pain as he fell

through the door, slamming it shut behind him, the heavy clunk of the security door locking him in.

The gunman made it to the building window, peeking in through the opaque security glass, but Ash thought he probably couldn't see anything.

A screeching sound came from the building: *the guard had set off an alarm!*

It bellowed, assaulting Ash's ears.

A distant police siren added to the din, bringing a blue light that illuminated the streets with each revolution.

Ash turned to the bridge; it offered an escape from this craziness. *But James Leigh is in trouble, and he can't leave him behind.* 'Come on, let's get James,' said Ash.

'Really?' said Steve, but followed, anyway.

They found James, incredibly still alive, making panicky, tiring efforts to climb out of the mud onto the raised pathway. Ash reached down, offering him a hand to grab, freeing him from his trap.

*Wasn't he shot? Where's the blood?* Thought Ash.

James, now on all fours, is breathing hard.

'Come on,' said Ash, scanning around for the gunman.

The killer, running back, saw them. He raised his gun and fired. They ducked in unison, Ash hearing the thud of a bullet into a nearby tree, causing chips of damp bark to fly.

They ran under the cover of trees, crossing the narrow metal pedestrian bridge, their clatter making it clear where they were.

Ash chose the road straight ahead up the hill; he knew it would bring them to a junction where they had a better chance of losing their chaser, the others, happy to follow.

Near the brim of the slope, Ash told himself they

couldn't afford to stop, but his lungs had other ideas; they were bursting. He's slippery with sweat, reduced to walking. They all are. Blood is pounding in his ears, feeling like he would be sick, and the cold air is hurting his lungs. He coughed, leaning against a wall, trying to stifle the sound with his arm.

Then Ash could hear the steady tramping sound of the killer on the bridge. *Coming for them.*

'Where to now?' asked Steve, tears forming. 'We're going to die, right?'

Ash pushed himself away from the wall. 'Over there,' said Ash.

He got under a tarp covering a caravan, dislodging gathered water and holding it aloft for the others to join him. Ash had watched thrillers on TV and was sure he could make better decisions than the characters who got themselves into trouble. But Ash had a new understanding of how fear could affect you and how making the right choice became less obvious. *He would not be so quick to judge those characters in future, if he had a future.*

Ash peeked under the tarp and light rain and could see the man appear at the end of the road. He came to an abrupt halt, looking around, trying to decide which way to choose.

*I'm an idiot. If he chooses this way then I've trapped us here. Do we run?* Thought Ash.

'L-look,' said Steve through his tears.

*Torrents of heavy, beautiful rain fell, making it difficult to see up the road–which meant it would be difficult for the killer to see them, too.*

It didn't stop Ash's imagination from playing tricks with him, making him see shadows in the rain. *His skin crawled, along with a feeling that he might just piss himself.*

After an age, the rain thinned–enough to see no sign of the man.

*Do they risk coming out? I'm scared, but I'm also really cold.* Memories remain fresh from his night out in the open after seeing Angelica. *He had to get a grip and make a decision.*

*'Come on,' he said.*

They emerged from under the tarp. Ash strained to hear the killer hunting them.

What he heard were Steve's teeth chattering. Ash gave Steve a sharp look, and Steve clamped a hand on his mouth to stop the betraying sound from escaping.

Ash's chest hurt from holding his breath, and he eased a breath out.

'I think he has gone,' Ash whispered.

'He fired at you. But did he hit you?' said Ash, directing his question to James.

'My side hurts,' James admitted.

He shoved his hand under his clothes and winced at the pain, but his hand returned without blood. He removed his shoulder bag, examined it, and found a hole. Inside the bag, his math textbook had taken a hit.

'At least the book was useful for something,' said Steve, wiping his tears, a little colour returning to his face.

'What? Oh yeah, funny,' said James, giving a tense smile.

They stood in their triangle, bonded by their escape.

'Why does he want to kill you?' asked Ash.

'I don't know.'

*Being a member of a big crime family might have something to do with it,* thought Ash.

'What do we do now?' asked Steve.

'Come by mine, it's not far,' offered James.

Ash didn't want to go to the Leigh's place. But he felt he couldn't wriggle out of it. Steve's nervous glance towards him signaled the same thought.

*The Leigh's place: There be dragons*

JAMES'S HOME is a 400-year-old Sussex thatched, detached house in the centre of town.

James's mother appears in the hall. To Ash, she had a bit of the dragon about her, too; her hair pulled taut like scales, her skin leathery, and her eyes cigarette smoke jaundiced, suspicious, seeing through him.

'Who are these two?' she asked, wary, eye contact unbroken with Ash.

'Friends, mum. Can I get them a hot drink? We're cold.'

'I suppose.'

A man, who Ash assumed to be Reg Leigh, walked into the kitchen.

*Here, definitely be a dragon, as if he could spit fire.* Curly thinning grey hair, fleshy face, green eyes squinting above heavy greyish bags. His chest is immense, like he'd stored away air for emergencies, his biceps standing out, reminding Ash of a well-stuffed sofa.

'Who are your friends, then?'

'Ash and Steve, they saved my life, Dad.'

'What do you mean?' asked James's mum, slipping an arm around James's neck, pulling him close, a little of the dragon melting away before Ash's eyes–until she looked back at him.

James told his version of the events to his parent's narrowing, calculating gaze. Then, they were shepherded into the lounge *so the adults could talk.*

But when they started talking, even behind the closed door, Ash could hear every word.

'An attempt to kill James? Would they dare?' asked James's mum. *Did she mean the Marshs?*

'We agreed to carve up the town with them to avoid unnecessary bloodshed, but Tommy's death has changed all that.'

'Did you have anything to do with Tommy's death, babe?'

'For the umpteenth time, no. Yes, times have been bloody hard, and the share of the pie is shrinking, but we don't need a war. But maybe the Marshs need to squeeze us or die.'

'Then they can fucking die, babe.'

'I spoke to Marsh, and he said he had nothing to do with it.'

'And you believed him?' she said, voice rising.

'Now hang on, love.'

'Don't *now hang on, love* me. Their boy lies in the morgue through no fault of ours, and then someone attacks our son. What about it, babe? How many more clues do you need? Are you two going to spoon after he has humped you?'

'Don't say that. You should trust me.'

'You can't be fooled, can you, babe? Let me tell you something, your pride in thinking you can detect the truth is a crock of shit.'

'What does that mean?' asked Leigh.

Ash thought *there was so much unsaid if you listened. And there seemed to be so much not being said in the Leigh's kitchen.*

'Listen to me, babe,' said James's mother. 'It's a shame their boy is gone, and they want someone to pay. But it won't be me, not with my boy.' Then added quickly, 'Our boy.'

Ash saw James's nervous glance in the kitchen door's direction, tight-lipped. *We shouldn't be hearing this; that's what James is thinking.*

James knocked on the door and entered the kitchen.

'Ash and Steve need to go.'

'Lads,' said Reg Leigh, joining them in the lounge. 'I am grateful to you for saving my boy and I won't forget it. If you need anything, let me know.' He slipped them fifty credits each. 'Keep this to ourselves, right?'

Ash felt intoxicated. He had never seen so much money in one go, and saying *no to the dragon wasn't an option,* he was sure.

# 8

# Holroyd

Holroyd is going through her exercise routine when the front door swung open, leaving a handle dent in the wall.

Grey entered, keeping his coat on.

'Not staying?' Holroyd asked.

He darted into the kitchen, filling the kettle.

'Not for long. Tea?' he said, words cut lean.

She gasped a 'Yes' in between squat reps, a half-moon of sweat around the neck of her scarlet red training top. She took a breather, asking, 'Is everything alright?'

'Yes.' Clipped. *Evidently not,* she surmised.

She could feel irritation welling at his obtuseness, but instead, she offered her latest update. 'I picked off the Butlers today. I've got compassionate leave from Fawngate because of Harry Butler's death, which means I can pull forward the other targets.'

'Good. The sooner we finish here, the better.' Holroyd could only pick out a hint of an East Midlands accent when Grey got stressed or drunk. She caught the accent now.

*What's happened? And what she had to say wouldn't improve his mood.*

'I've seen a boy in my class before.'

'Where?'

'At our last assignment. I caught him staring at me. Not in a lovelorn way, either, before you comment.' Not that Grey was in the mood for humour. 'I looked at his file at school. He was definitely a pupil at Warburn Secondary.'

'When we were there?'

'Correct.'

'What's the odds of a teenager crossing our path in two places? It's bad luck. But we can't afford for him to be putting two and two together.'

They worked well as a team. Only Grey was making mistakes, and Holroyd had her suspicions that he isn't telling her things. He'd also developed this habit of obsessively cleaning his gun. *Was their work getting to him?*

'Something on your mind?' He asked. Not looking up.

'What are you not telling me?'

'Straight to the point, as ever. I am grateful for your honesty.'

'Can you offer me a reason to be grateful for yours?'

She waited for a reply. Grey eyebrows furrowed, appearing to mull over his options.

He sighed; the decision made. 'I've fucked up. And I'm trying to fix it.'

*Good. There's stress in not knowing. Now, they could try to resolve it.*

'Okay. What have you *fucked up*?'

He laid out what happened at the trading estate: about killing the security guard and two witnesses who had slipped away. And finding James Leigh's body missing by the river.

*Grey is right–it is a mess.*

'We're a team. You should have told me before.'

'Yes, I know. But I didn't want to implicate you. If our contact finds out, he'll have me dealt with.'

They both knew that she would get the job of *dealing* with Grey.

'He has not given me the order to dispose of you.'

Grey scanned her, looking for hidden clues of intent.

'If he had, you'd be dead by now. Be sure of that, fuckwit.'

His stern demeanour cracked, and he laughed. 'I don't want you in the firing line if the shit hits the fan.'

He grabbed a cloth and started cleaning his gun again.

'Are you saying you care for me?' her words laced with mock longing, but masking, even to herself, feelings.

'If you've got a kill order, will you shoot me and have done with it?'

Holroyd gave a low seismic heave of her chest, exiting as a small laugh.

She said, 'When you've finished playing with your penis replica, could we work on getting you out of this mess? I'm going to take care of the teenager, Heath. Can you finish putting a line through James Leigh's name so we can ignite a war between the Leighs and the Marshs?'

'Sounds good to me.'

# 9

# DCI MacGillivray

MacGillivray blew into his hands as he followed Crozier towards the warehouse entrance. The crime scene is embroidered with yellow police tape tied to planted metal poles and police cars parked in the warehouse forecourt.

PC Hanbury met them, looking pale.

'Your first?' said MacGillivray.

Hanbury nodded—a short, worried head movement, as if admitting to feeling queasy would sound the death knell of his career.

'Don't worry, we've all been there.' Hanbury nodded gratefully, then quickly revisited the bushes to reacquaint himself with the remains of his breakfast. Crozier gave him a tissue to wipe his mouth and then Hanbury led them to the warehouse entrance.

The guard's body lay slumped against the wall next to the klaxon button. A delta of blood and rainwater had formed on the dusty concrete floor, grit offering a token resistance to its hybrid flow on the cold concrete floor.

'The security guard, Keith Bakewell, took two bullets, according to the forensic pathology team,' said Hanbury, 'The first shot looks like it hit him when he was in the security shed at the gates, going by the blood on the floor. The second, as he arrived at the warehouse. By the time we got into the building, he'd bled out.'

'Thank you, Hanbury.'

'Could it be a *Leigh* job?' suggested Crozier.

The Leighs are the closest rivals to the Marshs in the area, and MacGillivray thought the raid on the warehouse fitted Leigh's modus operandi: the ransacking of white goods factories, one of their signature enterprises, along with their prostitution racket and the drug territories shared between them and the Marshs.

'You think the guard Bakewell was in on it?' asked Crozier.

'One way to get past the guard is to buy him off.'

'But he *died*.'

'It's not the first time crooks have double-crossed each other.'

'The factory owner is organising an audit of stock, though, on an initial inspection, reckons nothing's missing.'

'A failed raid, then.'

Crozier looked unimpressed with MacGillivray's logic. *When she's in charge, she can make the calls.*

But now Crozier appeared preoccupied, gazing across the river.

'What's got your antennae up?'

'It feels like we're being watched.'

MacGillivray scrutinised beyond the stretch of water to the residential rows of terraced houses on a hill but couldn't discern anything unusual.

'A nosy neighbour?'

Crozier didn't look convinced, shaking her head, a signal to return to the topic.

'Anyway, don't you think it's odd that there's a cluster of innocuous cases? After so many violent crimes - and this case seems to be at odds with the recent trend – highlighter a statistical anomaly,' said Crozier.

*Before the internet crash, computers came with new techniques and a new language. Arming the bean counters with more easily gathered data. Despite the old non-tech ways returning, the bean counters stay with us - as does jargon like 'statistical anomaly'.*

'Could an *anomaly* also be a coincidence?' said MacGillivray, struggling not to sound peeved.

'Even if the victims are criminals?'

MacGillivray breathed. He had to admit that Crozier might have a point. He would like to double-check and follow up, so everything is squeaky clean. But he had a skeleton staff, an anaemic budget and precious little time–it's all just a firefight.

'Listen,' said MacGillivray through gritted teeth. 'The Leighs wanted access to the factory, but it had all gone wrong. And now we need to move on and not waste precious resources on a dead criminal.' MacGillivray's mind is made up, but he detests the impatience that filters through in his words, as well as the reproachful look he receives from Crozier.

'What's that?' asked Crozier, nodding towards a tree.

She guided him to a space of greenery between the warehouse and the river pedestrian bridge.

'Look. A fresh scar in the bark.'

He peered closer and saw the remnants of a half-buried bullet. The white, torn flesh of the tree peeking through.

'Well spotted,' said MacGillivray. *She is all vibrant energy*. In a moment of self-awareness, he considers that she might have the opposite opinion of him. Not long ago, he would've spotted that tree clue, too. And, dare he admit, be hungrier, keener.

'Why was someone firing in this direction?' posed Crozier.

'Perhaps the gun went off by mistake?'

'The killer made two hits from a distance - that's a professional shooter, one that would not shoot by accident.'

'Good points. So, the killer, or killers, were firing at someone else. That suggests that we have witnesses out there.'

'Can you get onto forensics, Hanbury? See if the bullets came from the same gun.'

'On it.'

'Let's speak to the guard's wife and see if we can ascertain if he was in on the job,' said MacGillivray.

The last departing vehicles left the industrial estate, brake lights lighting up exhaust fumes, headlamps projecting funnels of lights into a wet, tenebrous sky.

## 10

# Holroyd

Holroyd hated rushing a kill. She always wanted her planning to be as meticulous as possible.

Grey's trip-ups, added to Ash Heath seeing her at the last school, meant she had little option but to cut corners. Her one concession was to wait for the day to darken, using this grim November weather as murder-hiding cover. The rain had fallen for hours; where *did it all come from?*

Holroyd found the Heath boy's home. *It's hard to believe that a severe threat to Project Deadhead lived here—a two-up, two-down terraced house in a less-than-salubrious part of town.* The front door had layers of brown paint peeling off. Broken slatted wooden fences peaked through the stacked refuse sacks like bad needle teeth. A couple of motorbikes and a car sat in the drive under an awning, maintenance projects waiting to be finished. The dust and dirt of the hot months could cover a multitude of sins. It was a one-size-fits-all film of thick dirt, hiding people's embarrassment, their lack of pride and hope, until the rain exposed it again. And Holroyd is going to leave them lacking their son.

The Heaths lived opposite a park, denuded of

trees, cut down by people who could no longer afford coal or gas heating, so they raided the park for anything to burn. She remembered her first trip to a park in London as a child, looking up at the trees, gazing in amazement at its colours and beauty. She felt a surprising hurt from the brutal crimes against nature in the name of survival.

Holroyd hid behind a park metal fence, invisible as a careful abuser's bruise.

She saw Ash Heath walking down the street alone. She barely felt a rise in her heart rate. *It's going to be too easy.* Holroyd could intercept him by the garages and deal with him, out of sight. Not that anyone was around. The street had an air of abandonment, except for a trail of cigarette smoke rising from a Merc's red two-seater open window. *Perhaps I'm not as alone as I'd like to be.*

The car's engine started. The driver set off, slewing its way towards the boy.

Holroyd swiftly retrieved her gun from her shoulder holster.

A man got out, holding a knife.

*Perhaps my job's going to be done for me.*

# 11

# Ash

*I'm almost home,* Ash thought.

Since parting company with Steve at his friend's home, he'd felt even more exposed. In the last few hours, there had become so much more to fear: The Marshs, the mystery killer, and the police - for questions he did not want to answer. And Reg Leigh is supposed to be a friend, so why did he feel so *threatened* by him, too?

And with darkness falling early, sounds took on hidden depths.

Though there's nothing hidden about the sound of a motor engine racing up the street, breaking, rainwater splashed aggressively against him.

'Hey!' Ash shouted, jumping back.

The man in the red sports car looked familiar. *An older version of Tommy, maybe?*

The man got out of the car, holding something sharp and nasty. He waved the knife threateningly, manoeuvring Ash into the space between the garages. Ash tripped, falling backwards into a dead-end harbouring a soggy pile of leaves. He got to his feet, looking frantically around, but could not see an escape.

'No, please,' he pleaded, hands held flat out, shields of futile flesh.

'You killed Tommy. I don't do fancy speeches, so let's just get it done.'

'No, wait, I didn't. But I saw the man who did.' His words burst from him, supercharged with fear.

'You're lying.'

'Honestly. The same man tried to kill James Leigh today.'

'You're just making up shit to save your skin.'

But the knife, Ash noted, had stopped moving towards him.

'What did this *man* look like?'

'He's...'

'Troy!'

'What the hell?' hissed the man with the knife, turning.

*It's Troy Marsh? Tommy's brother? And he wants to kill me!*

Reg Leigh stood on the corner, gun in hand. Two heavy-looking men are next to him.

'I don't want trouble. I only want to talk to the boy,' said Troy.

'Neither do we. But you can't have the kid. Leave now, and we'll avoid any unnecessary problems.'

Ash watched Troy look at the three people; he seemed to think they were too much to take on his own because he backtracked to his car.

'I won't forget this,' said Troy.

Ash isn't sure if Troy is talking to him or the men. Troy Marsh got into his car, slammed the door, then drove off, cutting a wheel-spinning swathing through the water.

Even though relief washed through him, Ash felt

pressure in his bladder, as if he might wet himself. *Please, no,* he pleaded with himself.

Reg Leigh crossed the street towards Ash.

'The Police are looking for you, Ash. You won't say anything, right?' said Reg Leigh.

Ash, pale, feeling sick, said, 'No.'

'Good lad. I said I'd look after you, didn't I?' He walked away, the two other men falling into step behind him towards his car. He didn't feel looked after – he felt more like he'd been used as bait.

And Reg Leigh still looked like a dragon.

## 12

# DCI MacGillivray

MacGillivray and Crozier arrived outside the Bakewell family home, a two-bedroom flat in one of the sad looking sixties built high-rise blocks in the centre of town.

The rain relented, teasing with a promise that it might stop. MacGillivray looked up at the block. He noticed a single jet stream passing through a gap in the clouds. The time was when the skies were scarred with vapour trails, like myriad skaters' trails on ice. Now, most airports lay in desolate, economic ruin. People used to say planes shrunk the world. *The world must have expanded again because hardly anyone was going anywhere now.*

The broken lift meant a stair hike to the sixth floor. *Lovely,* he thought.

With barely elevated breathing, Crozier is waiting for him, arms folded theatrically. MacGillivray laboured his ascent of the stairs, stopping periodically to gulp down air.

'Someone needs to get their daily steps in,' said Crozier.

*Again, she was right. He needed to sort himself out.*

A man in his early sixties, short, thin, hair thinning grey, wielded a brush, moving pools of rain under the railing, generously shouting a warning to those below.

He eyed them, then said. 'The Bakewells are good people,' grimacing with the realisation that the adults in the house were reduced from plural to singular. He'd quickly labelled MacGillivray and Crozier as police, guessing why they're here. 'Why don't you use your time to catch real criminals?'

'I am DCI MacGillivray, and this is DI Crozier; if you have anything useful to tell, then call me,' said MacGillivray, handing him a business card.

The man turned away, mouth downturned, to continue sweeping.

Crozier led the way to flat six-zero-four. MacGillivray followed, massaging the *v* of his thumb, unsatisfactorily.

'HOW AM I?' Lauren Bakewell said, baby in arms, tears gathering at the corners of her eyes. 'What do you do when your husband has been shot dead? When he's suspected of stealing from the factory and his good name is being trashed?'

Shored-up grief sat behind the stern face of Mrs Bakewell; a nerve flickered in the bag under her right eye.

'Could we discuss this indoors? You're letting the heat out,' said Crozier, placatingly.

Mrs Bakewell turned around and walked down the hallway, leaving the door open. They followed her in, accepting the implied invitation to enter.

A degree and a master's in psychology hung on the hall wall. Laura Everson's name on them, presumably her maiden name. MacGillivray hoped her skills were not going to waste, though society didn't seem to value the healing of minds anymore.

'What do you do the day after?' She spoke through clenched teeth, jawline resolute. 'And the day after that? Having children is a distraction from the pain of what happened. Yet the unquantifiable in-tray of grief waits,' she hit the words home, stubborn nails into knotted wood.

MacGillivray opened his mouth to speak but hesitated as he saw her resolve cracking as she fought to hold her inner turmoil at bay. *Something he could identify with.*

'I know your question is a piece of social etiquette, but please, get to your point.'

*Isn't this what we all did, oiling the cogs of communication? But her words are a damning inditement. He's so out of his depth in this part of the job, and she's got the skills to tell.*

*And then there's this fear of his: that a lid is being levered off his own crypt of feelings. And the fear of finding something within.*

'I need you to catch the swine that did it and clear my husband's name.'

Her words are like questioning splinters into MacGillivray's mind.

*Is he doing enough for her? Had he already, all too easily, condemned her husband?*

He reached subconsciously to scratch the skin behind his ear—the sticky dampness of his weeping dermis staining his nails.

'I feel like crap, okay?' she said more softly. Her two-year-old looked up. Distracted from playing with brightly coloured blocks on the floor, she said *crap* and laughed.

Mrs Bakewell stopped for a moment, drawing in a

breath. 'Don't use that word, Alice; Mummy is naughty for saying it.'

'Naughty Mummy,' said Alice, her smile a blossoming rainbow.

Lauren Bakewell switched looks between MacGillivray and Crozier. 'What are you going to do?'

MacGillivray bit the bullet.

'Mrs Bakewell, I am sorry, but I must ask. Do you know if your husband had dealings with Sam Marsh or Reg Leigh?'

Mrs Bakewell's sour laugh quickly switched to anger.

'Look around you. Are we *on the take*?'

MacGillivray couldn't deny the scarcity of furnishings, of clean but ageing decor.

'We're not suggesting...'

'Yes, you are.'

*Yes, I am.*

She stomped to a bookshelf and pulled out a bright yellow folder. She laid it on a table and opened it, exposing bank statements–copious minuses taking the account into an overdraft.

'Look. A record of debt or another clever cover-up? Don't worry about a warrant; go ahead. Search the flat; you'll not find a stash of *cash, gold bullion, diamonds*, or *hot merchandise*. Go on. Is that enough of an answer? Do you have any more questions? If not, then could you please leave?'

'We're going to take you up on your offer and search.' MacGillivray said to the resigned shrug of their reluctant host.

HAVING FOUND NOTHING, they ran the gauntlet of Mrs Bakewell's stare.

'Happy now?' she said.

MacGillivray could not deny the sincerity of her belief in her husband. If she's pretending, she's a talented actor.

Lauren Bakewell hugged herself as if trying to hold herself together, eyeing him.

She said, 'DCI MacGillivray, you fool yourself into thinking you're coping. But look at yourself; your body does not lie. Distractions and addictions are your crutches, aren't they? You need help because you aren't facing up to your trauma. Whatever it is.'

He stared at her as if she had handed him a life sentence.

Suddenly dry-mouthed, he quickly found the words to exit on, to escape. 'Please call me if anything occurs to you, Mrs Bakewell.'

They left the frostiness of the house for the cold, more comparatively comfortable concrete walkway.

'She's got skills,' said Crozier. 'The only thing missing's a psychiatrist's couch.'

'Grappig, Crozier.' *Except it isn't funny. He's shaken.* Lauren Bakewell had rattled the gates of his dark fortress of grief that he'd defended for so long - and she'd freed up something he still isn't ready to face.

'He might be innocent,' MacGillivray said.

'Really?' said Crozier.

---

DIRT, pooled water and leaves obscured the parking bay lines in the police station forecourt, so MacGillivray guess parked.

Crozier got out and walked to the entrance. He caught her furtive, concerned glance back at him. He nodded at her to go on. He had beaten himself up for so long with the guilt of how he had let his family down, kept his grief bouncing against the steel wall of his inner fortress. Yet Lauren Bakewell had cracked his defences with her pride in her husband and for the risks he took to support his family.

MacGillivray's wife had believed in him, too. She had stood by his values, of what it meant to be a police officer. *She had been proud of him.* Maybe he'd not wholly comprehended that she'd understood the risks to *all* of them.

His wife had once asked him, 'What if I put myself in harm's way, in my job?'

'I wouldn't want that.'

'Have I got this right? You can put yourself at risk daily. Look at me, the big brave police officer, going into danger—the sheriff of the town. And women can't be allowed to face danger. You're an anachronistic, chauvinistic git!'

He had cringed at the abrasive truth.

And then she laughed. He laughed, too. She understood, beneath it all, he didn't want her hurt. But it didn't make him right.

'Now you've been shown the error of your ways, sheriff; give me a kiss, get yourself on your horse, and go.'

HE ALLOWED himself a begrudging smile at the memory and of the last time he had truly laughed—the view out of the windscreen blurred.

He isn't religious. *But if there is an afterlife, how would his wife and daughter view him? Are they nearby watching him? It's strange, but they seem closer to him now than when they were alive. When he had chosen work, choosing to neglect his family. And wasn't it ironic that he hadn't been so present in his job since their deaths?*

He put his head against the steering wheel. Tears running down his face, gathering at his chin, dripping onto the steering column and his trousers. His sobs exercising muscles not used since childhood, when crying came easier. It hurt to feel his toxic masculine shame burn at losing control.

He opened the door, bent down and grabbed a handful of cold rainwater. Splashing it on him, rubbing it harshly into his face as if it held the panacea for all his ills. To his surprise, he felt better. He hurt from his loss, but could he see beyond the darkly crenelated parapet of his inner stronghold? *Was there hope?*

And what's happening in Hawham churned in the recesses of his mind. His intuition spoke to him; the voice he had ignored was now pointing out inconsistencies. Making him challenge the conveniently packaged cases he had closed, filed under the excuses of *not enough budget* and *pressure from above.*

He had to drill down, examine, poke and prod to find something closer to the truth.

*We're a town close to the brink.*

*We, the Police, have few resources left; all we do is firefight.*

*But if not us, who is going to fight for justice?*

*If not me, then who?*

*Could the white flag he had been flying since the death of his family be furled and filed under 'redundant'?*

*For someone good at asking questions, he must step up when it comes to giving answers.*

# 13

# Ash

'Come and see me. My parents are out,' said Angelica on the phone to Ash, words coded with promise.

'I really want to, but the weather's terrible.' But it isn't just that.

He and Steve had seen Tommy *die,* and there had been Eddie's body, too. They'd followed the killer to his home. And James had been saved by them only to have nearly caught by the killer. Then he'd almost got knifed to death by Troy Marsh. Ash feels exhausted, not only physically, but he can hardly think straight, too, with these highlights of horror playing constantly through his head. He'd heard about *trauma.* He supposed he must have it.

'Yes, I know you *want* to,' said Angelica, her words like the screeching of tyres before an accident, scything through his thoughts.

Hard rain lashed against the windows; he could almost be at sea. Three miles is a long way to walk to her place. And he'd have to come back, too.

'Don't you like me...are you losing interest already? Maybe I need a new boyfriend.'

His breath halted in his chest, realising he didn't know what she meant to him, but he was not prepared to lose her, to lose this *excitement*. And she'd called him her *boyfriend*.

He remembered standing on the top board at the swimming pool. His mates had queued behind him, egging him on. He'd looked down, trying to gather the courage to take his first-ever jump. The podium had felt like it was getting higher. And he'd bottled it. Having to take the walk of shame, burnt with derision from the others.

He didn't want to feel like that again; he wanted to take opportunities, take risks and be brave.

*And Angelica is an opportunity. She's stunning. Her family has money; they must do to live in a big house out there. He's punching well above his weight.*

But he also felt under pressure. And, deep down, he had the first stirrings of dislike for her. *But then, if she gave out, did it matter? But what did that make him? To treat her that way?* But what *could* happen had taken a grip, this feeling pulling him along as if he didn't have a say.

He could hear his dad telling him not to blame everyone else and to take responsibility—*more reason to feel bad.*

'helloooo,' said Angelica, pulling back to the now.

'You wouldn't, would you?' He said, dread lacing his words.

'I'll see you soon,' Angelica said, putting the phone down.

HE DRESSED for the weather and headed out the door, immediately feeling the sting of the rain, pulling his scarf tighter to his face. He left run-down suburbia, now walking down a country lane, its covering of stark trees a canyon of hollowed-out darkness, of this cloud-stuffed night, its protection scant against the rain that is turning to hail; small balls of hard ice breaking on the ground all around him. His hands and feet turning cold.

HE ESTIMATED he's covered two miles.

*It's strange to be miles inland from the sea and yet be able to smell the salt in the rain.* If he closed his eyes, he could imagine being back home, feeling the sea spray as he had walked on the promenade. *He missed his old home.*

He crossed the stone bridge, taking him over a swelling river. Water pushed at the effrontery of the bridge's columns, which opposed its progress. He swore he felt stones crack, panic making him run to the other side.

*It's so cold, so wet and so dark. I should've eaten something hot before I came out.*

*This is rough.* He should go back. But he wouldn't. *He could die because of sex. If he was to die tonight, better to die after than before, at least.* He almost laughed.

At last, he saw her place, rain defusing the beckoning light of Angelica's detached Victorian three-story home with a gated entrance. The luxury of a porch light shone across the driveway, *a two-fingered answer to the government's pleas to save electricity.* He opened the gate and stepped through.

The door is opened, and it's all Ash can do to stop his jaw from dropping at Angelica's beauty.

Angelica smiled, pulling him in. After he removed his dripping coat, she hugged him.

Her bright smile made him feel good; it was good that she's pleased to see him. She led him through to the lounge. It had that deep carpet quiet. Nice modern furniture and the warmth and smoky smell of a fire. He warmed himself against the luxury of the burning logs, its guard holding back sparking embers, the flame's gold light flickering in her blonde locks.

Angelica slipped her arms around his neck and leaned her forehead against his.

Her gold pendant hung against the softness of her neck, dropping into shadow. He could feel her breath, hot, sending shivers down his spine.

Her perfume hid a slight tang of something behind the citrus, reminding him of the wolf, again. *What warning is he ignoring? Did he care?*

'Beer?' she said, disengaging from him.

He nodded, feeling the blood from his brain moving south. He sat down, hoping to hide his embarrassment.

Angelica left a knowing smirk out there. She took a bottle from the fridge and opened the top with practised, intimidating ease, took a sip and offered him the bottle. The beer had an iced neck, cold burning his fingers. He had little experience of *drinking*. His parents rarely had alcohol at home. No shop would sell it to him, even if he had the money. And he'd not been keen on the taste when he drank it. Or the feeling of losing control. But he took the bottle from Angelica anyway, because he wanted to *impress*, to look *experienced*. He took a clumsy swig. It was, as he remembered, nastily bitter. Angelica held her hand out for him to return it.

She had another casual pull.

'Will your parents know you've been drinking?' cursing himself for saying such an uncool thing.

'My parents won't miss the odd bottle,' she said.

Despite himself, he was impressed.

She moved towards him, her arms pulling him

towards her. Her mouth is slightly open, inviting. Her sweet breath tightening his stomach.

'Do you like me?'

She ran a hand down his back, moving around the edge of his belt.

'Yes,' he croaked, throat suddenly dry, swallowing suddenly a ridiculously difficult thing to do.

'Would you do something for me, Ash?'

*He would do anything right now.*

'What?'

'It's important to me,' she added airily. 'The cow in Mapleburn Manor has this baby crib her firstborn died in. It's all she has left of the child. I want you to take it. I hate her.' As if that explained everything.

*He'd had many thoughts about Angelica, but being 'ugly' is a surprising new one. And here's the wolf, the thing to be wary of.*

'Why?'

'She saw me in town drinking and told my father. He grounded me for a week. I think he was more annoyed that I was with a *man*.'

A *man?* He felt threatened, intimidated and confused.

'I'm not sure,' he said weakly.

Angelica got to her feet, primly brushing the front of her skirt with her hands, causing the exciting warmth to evaporate into the vacuum of a cold, dark, non-Angelica space.

She denied him eye contact while raising her beautiful blonde eyebrows.

'I have a migraine coming on. You should leave now.'

Ash shivered, even in front of the fire. He'd never heard of a migraine. But he could tell it isn't good news.

'I might call you later in the week. Don't call me,' she said with cold finality.

'Wait, I didn't say *no* exactly. I need to get used to the idea.'

'Let me know soon. Before someone else has *got used to the idea* quicker.'

He mentally collapsed: wanting Angelica, wanting her attention, wanting everything.

'Okay. I'll do it.' Feeling nauseous.

'Good. I would have been *so* disappointed, Ash.'

'What should I do with it? The crib.'

'Lose it somewhere. I don't care. Just get it.'

'How will I get in?'

'The fools leave a key under the stone next to the entrance to the tennis court. And they're going to London tomorrow. So, it will be easy.'

He swallowed. 'So, Angelica...'

'So, Ash. Let me know when you have done the deed. I'm going away for a few days. Then, I cannot be blamed. But she will know it's my doing. Which, of course, is the whole point.'

'But...'

'Goodnight, Ash.' She took him by the hand to the door, kissed him on the cheek. Smoothly opening the door and easing him out.

ASH, bewildered, walked to the gate. The harsh rain prodded him, reminding him of its presence, snapping him out of Angelica's spell.

*The thing she asked. He couldn't do it.*

Ash turned back to her home.

ANGELICA OPENED THE DOOR, looking irritated.

'I'm sorry, I can't do it. And I won't be seeing you again. Sorry,' said Ash quickly before he could change his mind. He turned away from the now surprised-looking girl.

'Ash?' she retorted, a flaw of hesitancy cracking her haughty demeanour.

He felt relieved. But he also had a sense of losing something, feeling hollow.

Then he saw someone climbing over the wall.

'HAVE YOU CHANGED YOUR MIND?' said Angelica, sounding triumphant, as he darkened her door.

'You've got visitors.'

Angelica sneaked a look past the porch and saw the invaders.

'Oh, hell! Get in quick.'

Angelica rushed across the room, turning the lights off. Luckily, the thick bay window curtains had already blotted out the light. She pulled back one to get a slit of a view.

Ash stood beside her.

'They're coming,' she said.

She seized the beer bottle from the coffee table, emptying its contents into the kitchen sink, the air thick with the smell of hops. Then, deftly cracked it, the neck of the bottle dropping loudly into the steel basin.

The double-glazed windowpanes cracked. Shards of glass hit the curtains' lining, falling to the floor. An armed snaked in, a hand trying to unlock the window latch.

Angelica brought down the broken bottle, running its jagged edge along the intruder's hand. A male voice yelped in surprise, pulling their hand back, sustaining more damage on the fractured glass of the pane, going by the yelps.

'What happened?' said a voice outside.

'I've cut my hand.'

'You won't be playing the piano for a while.'

'I don't play the piano.'

'It's a joke, okay?'

'Yeah, fucking funny, prick.'

'Yeah? You're the prick for cutting yourself.'

'You need to keep pressure on it to stop the blood. Here, use my scarf,' said a third person.

'Thanks, James. Now, let me get this window open.'

'Don't use names, idiot.'

'Eh, yeah, sorry.'

Ash recognised the voice. *That's James Leigh.*

'Oh, my fucking hand. There's someone inside.'

'What? They are supposed to be out.'

'Well, they're obviously not. Jeez, that hurts.'

Angelica gripped Ash's hand, pulling him away. They took the staircase, taking double steps in the semi-dark. Ash stumbled but regained his balance. Ash heard angry voices from the lounge - *they were in the house!*

Angelica pulled a gun from behind a thin cabinet.

*'You've got a gun?'*

'My father keeps one hidden here for such occasions,' Angelica whispered.

*How often does this happen?*

'Shut your mouth, Ash, it's not a good look,' she said as she loaded cartridges into the weapon, the weapon cracking loudly.

*She's so calm.*

Angelica reached for a pole, which had a hook on the end, using it to bring a loft ladder down.

'Up there, go,' she ordered. Her momentum shunts his fear into a temporary siding.

Ash climbed, turned at the top of the ladder, and reached down to help her. He could make out silhouettes in the hall and a flash of harsh gun metal.

A member of their group found a light switch, making light erupt, and its effect was blinding.

Three stood there, masked.

One of the intruders fired their shotgun, hitting the top newel post on the landing staircase, causing wood shards to fly. Angelica, halfway up the ladder, fell, her gun scattering across the landing. Ash impulsively scrambled down, picked up the weapon and stood between Angelica and the intruders. He levelled the gun but is unsure if he can pull the trigger.

The smallest of the three locked eyes with him, widening in recognition. *It* is *James.*

James put a hand on the gunman's arm. 'We're leaving,' he said.

'Why?'

'Because the old man will kill you if you hurt him.'

The man lowered his weapon.

'They haven't seen us. No one is hurt.'

'Speak for yourself,' said the man with the cut hand.

'What about the damage to the staircase?' said Angelica.

Ash thought she was pushing her luck. *But so coolly.*

'Count yourself lucky, so don't try anything stupid.' Then they left.

. . .

ASH AND ANGELICA slumped down onto the staircase.

'One of them knew you, didn't he?' asked Angelica, not bothering to wait for him to confirm it. 'Who are they?'

Ash would not say. And he didn't want trouble with the Leighs, either. *Isn't James his friend now, anyway? It seemed that way.*

He ignored Angelica's question. 'I'm ready for a beer,' he said. Remembering his manners, he said, 'Please.'

'Good idea.'

She picked another bottle of cold beer and opened it, this time passing it straight to him. His hands shook, the beer an instant anaesthetic hit–*maybe it isn't so bad after all.* He passed it back; she took a glug.

'You did okay. Most boys fill their trousers if my father turns up, much less deal with something like this. And you jumped down off the ladder to protect me.'

Ash shrugged. 'You've had burglaries before?'

She nodded. 'And it will happen again.' She scrutinized him and then said. 'I think we've got off to a bad start. You don't have to do that thing I asked for.' Her hands clenched in front as she delivered her peace offering.

It was tempting. But he'd gone from being mesmerised by her to being wary of her, his trust broken.

'No, sorry, we're done.'

'I've messed up, haven't I?' She looked crumpled. *Was it another act?*

Even so, he asked, 'Do you want me to stay until your folks get home?'

'No, I'll be okay. They won't be pleased to see me alone with a boy here, on top of everything else. You were not here, you understand? I'll say I was spooked by a bird that had got in.'

'What about the gunshot damage?'

'I'll say I did it. Let me know if you change your mind. Will you be alright getting back?' she asked.

'I'll be fine.' The world had changed. He didn't have a girlfriend. But he felt better, *cleaner*. And he'd done something brave–*maybe I'm not as big a coward as I thought.*

He stepped back out into the night. The rain, if it's possible, fell harder. *It's been a strange day. And it's not over yet.* He faced the walk back home.

THE ROAD IS FLOODED, with water spilling over from the river, pooling into the dark, unknown depths. He didn't fancy getting his feet soaked when he still had a few miles to go.

He heard gunfire; Angelica's preparation to face her parents had begun.

*He had to bypass the flood.* Thorny bushes on one side of the road dropped into the river. It swirled, agitated, too close for comfort. He saw a dark coppice on the other side. *Get around it and he might get past the flooding.* A low barbed wire fence stood before him. Mind made up, he carefully climbed it at a point where a post and angles of timber held it firm, where he could gain a purchase.

It quickly became hard work, wading through the muddy, waterlogged field. He made slower progress as the ground rose. He'd become disoriented in the heavy rain, in the pitch-black night. Water had got into his boots, feet now miserably wet. He made it to the cover of the trees, now unable to see his hand in front of his face. Water fell in big gobbets from the trees, scoring regular, dispiriting hits on him.

His face is so cold; his lips are numb. He pulled his hood down tighter, breathing into his scarf, pulling toggles to close off the world. He almost wished for the stifling weather of not long ago.

The white noise of the rain settled upon him. He followed the path, leaving the trees, feeling like he's departing a spaceship to walk out into the void of space. After a while, he had to face it: he was lost, his inner compass failing him. He could feel the rain seeping into his so-called waterproof coat–*what a piece of shit*; mimicking a phrase Steve would use.

*It's got to be simple: if lost, retrace your steps.* But Ash had lost track of where *behind* was. Trickles of fearful sweat ran down his back as he fought to move in cloying, energy-draining mud. *Maybe he was more tired than he thought. He'd already been exhausted before he came out tonight. What should he do? If only someone were around to decide for him.*

The rain kept falling, soaking him, mingling with the prickly heat of scared tears.

Ash felt like lying down and resting for a couple of minutes. He had never felt so tired, and he just wanted his bed, but he knew he shouldn't risk falling asleep out here in this cold. *But a minute wouldn't hurt, would it?*

The rain slowed because he had made it under an oak tree and could feel soft padding on relatively dry ground. When he sat down, leaning against the yielding bark, he felt a little better. *He'd rest his eyes, just for a moment.*

# 14

# Holroyd

A single chime from the church signalled One am.

It's dark, quiet. The rain had abated. Her furtive progress through the streets of suburban Hawham so far unnoticed. This area of town is grim, semi-industrialised, a commercial place. It carries a hallmark of confused town planning, mashed into a backdrop of ageing depression, waiting for the wrecking ball to arc into bricks which had stood for an age, ready to give in behind a cloud of dust. But she guessed that a lack of development money had delayed an appointment with the demolition crew.

All but two rebellious, rain-defying young had given up on the night. Warmed by the illusion of heat that alcohol and desire bring, they were copulating in a rusting corrugated metal bus shelter, oblivious to Holroyd in their passion.

*If they don't notice me, they live.*

Thirty minutes later, she arrived at the outskirts, where the middle classes, or those with aspirations to that station, lived.

The council's night street lighting switch-off made her job easier for moving around unseen, though harder for spotting people.

Without light pollution, the stars took advantage of a break in the clouds, catching her off guard with their joy. She fancied she could float off the planet towards the bright, ancient points of light.

She sensed a presence coming up behind her from across the road.

'What do you see?' a stranger asked.

She faced her questioner as he appeared from the shadows of trees. A large man in his twenties, she guessed.

*Can a man walk as if he thinks he owns you?*

'Sorry?' Holroyd said dismissively. Feeling her balance, instinctual preparation.

'Do you see your future up there in the stars?' His head tilted forward, carrying ominous purpose and a faint contrary smile. 'Hand over anything of worth, or there won't be a future for you to see.'

His smile graduated to an assured, cruel grin.

'That's long-winded as threats go–you'll have to put some work into that,' said Holroyd.

Any vestiges of humour vacated his face. Lines of anger, new tenants. 'Hand over your fucking money and jewellery, or your days are over. And I may have *fun* before I'm finished with you.'

'Don't you have trouble finding it at the best of times without digging it out in this weather?'

It took a few seconds for her meaning to sink in.

'Do you have a death wish?' he spat, fumbling for something in his green Parker pocket. His knife gathered starlight, drawing it like a magnet, accentuating the edge.

Holroyd said, 'Only for men that feel entitled to engage in violence against women.' She jumped him before he could react.

Holroyd grasped his forearm, snapped his bone in

a single twist. The man's knife skittered into the watery groove of the gutter while her would-be assailant howled in startled, agonised pain. Holroyd's knee made brutal contact with his genitalia, causing a rapid exhalation of air from his lungs. She spun him to the floor; her palm pulled back, then punched down hard, breaking his teeth.

'You 'ucking *'unt.'* The downed man said through tears of pain and altered articulation.

'Is that acceptable language?'

She stepped back and swung a kick, the side of his head taking the brunt of the impact.

She could have walked on, spared him. But the problem is that he had seen her face.

Holroyd retrieved the dropped knife and moved towards him, ignoring his squeals of pain and gratuitous curses.

His look of undiluted terror meant he now truly believed she could wield the weapon without regret. His shock at the reversal of roles comical.

She remembered his threat of *fun.*

*Hadn't she been told, by the powers that be, to avoid extra killings? But if anyone deserved to be on the kill list, this piece of excrement did.*

She bypassed the defence of his one good arm and severed the primary artery in his neck in an efficient motion. *Blood's thicker than water, as the saying goes. But water can do a good job of thinning it.* She watched his red fluid swept away by the rainwater into a drain.

'What do *you* see?' she asked in a hushed tone.

She left him spending his last moments choking —the stars for company.

. . .

HOLROYD STOOD outside Leigh's place, a large, detached property with multiple vehicles parked in the expansive driveway.

*If people are going to break into a home, they'll succeed. But some security can make it difficult enough to question the worth of the risk. And this place only has one lock and no alarm. Perhaps Leigh imagined his status as a feared criminal would be enough to keep people out. Well, that belief has backfired tonight.*

Holroyd headed down the hall, a crack of light emanating from the lounge room. She stepped into the room. There sat the man himself, Reg Leigh.

'I do my best thinking in the middle of the night,' he said, utterly unphased.

He poured himself a whisky, perhaps not his first. Ice cubes turned somersaults in the small sea of amber, ringing a dull peal of bells against the sides of the crystal-cut glass.

'Can I get you one?'

The peaty smell permeated her senses.

'No.'

He took a sip, gauging her. Then said, 'No one can always be happy, I think. But isn't life just islands of happiness that we travel between? For me, the quiet of the night and a whisky - this is one of those islands.'

*Until her intrusion, of course. He's a cool one, she'll give him that, trying to distract her, playing for time.*

'So, Pendle wants to finish me. The double-crossing bastard.'

'What do you mean?' She hadn't expected to hear Pendle's name. *What's going on?*

'All in good time. You know my activities have brought me wealth. But here's the thing: success has never given me what I crave: a rise from my working-class roots into the upper echelons of society; it's the same old boring story of a lack of acceptance, of self-worth. Apart from my wife, you're the only other person I've told this to.'

*An ominous statement–denoting that he is confident that I will not live to spread that information around.*

Despite Holroyd's need to get on, she still asked him, 'What did she say?'

'Why not be happy with what you have? Ignore those stuck-up sods.'

'She sounds wise, if a little coarse.'

*Leigh's inferiority complex drove him to be something 'better' by any means. So what? There'll be no exoneration from her.*

She was itching to use her gun. *But if he is talking, there's a chance he might say something useful.*

'Money has only taken me so far because, I admit, I have tried to buy my way into the revered local golf rotary clubs. But they have standards I can't understand, much less meet. And with the baggage and culture of my upbringing weighing on me, I'm unable to send the right social signals for access to this world of entitlement.' He paused.

*Had he been ingesting too many do-it-yourself therapy articles?*

'I don't want to live where I started out, to be reminded of what I am. But then I hate living here on the *better* side of town. Being reminded of what I'm not.'

'Where's this leading to? What's your connection with Pendle?' said Holroyd, beginning to tire of his diatribe.

'Bear with me a little longer. I held a grudging

respect for their entry standards, those so-called elite; their acceptance held value for me. Now, with the economy and the climate going from bad to worse. And with the protection the police afforded the local privileged people becoming rare, suddenly, I'm someone whose *services* they need.'

He took another considered sip.

'What of their high-minded morality now when I can provide material things gained at someone else's cost? To provide security to protect themselves and their property. What of their newfound willingness to include me?

Once I had that bit of recognition, to be *seen*, I found its worth reducing–*the depreciation of satisfaction* - as I coined it.

My point is, if you kill me, many people will be unhappy. Because that is what you are here for: to kill me, right?'

He took another sip of his poison of choice.

She drew a breath, but before she could react, he said,

'Did you have anything to do with the attack on my son?' *A deliberate dropping of a distraction bomb.* Then Leigh, thinking her at her most unguarded, went for his hidden gun.

*A shade obvious.*

Two red holes blossomed in his chest as Holroyd's silenced bullets struck home.

She rolled Leigh over. He had a look of peace she thought he ill-deserved.

'No, me,' she said, 'But someone I know,' belatedly answering his question.

Holroyd made her way soundlessly up the carpeted stairs.

The elder son, Vaughan, who someone in her file had described as the *muscle without memory*, had used his last spliff; the stub had petered out in the bedroom ashtray. Now, Vaughan is dead to the world after Holroyd's bullet entered his brain.

In the main bedroom, Ally Leigh lay in bed, carrying the fight lines of her life across her forehead. She seemed to hold herself in tension, as if she fought even in her dreams. She stirred, alerted to Holroyd's approach by animal instinct. Her eyes now feral at the sight of the stranger.

Holroyd admired that.

Ally Leigh sprung at Holroyd. *Impressively so, from a lying position.* Holroyd's bullet entered her body but didn't stop her from hitting the gun free from Holroyd's hand. Then, her hands found their way around Holroyd's neck. Choking but weakening: Holroyd's shot had hit something vital.

Holroyd flipped Leigh onto her back and straddled her body, pinning her down. She picked up a warm, still head-indented pillow from the bed, forcing it across the woman's face. Holroyd's strength, the weight of her body, and Leigh's blood loss enough to do the job. Holroyd held the pillow down until she is sure the woman's fight was done.

She grabbed her gun, berating herself for making this job sloppy.

Holroyd checked the rest of the house and found it disappointingly clear. *Where's James Leigh? Grey had failed to kill him. Now he's missing here, too. How lucky could the boy be?*

She turned on the house gas. A tragic fire: a failed boiler. It's almost anti-climatically easy, as mass killings go.

She tugged at her black leather gloves and closed

the main door behind her, leaving a house full of the departed and a timer to spark.

Two minutes later, flames shot up, lighting the sky, and she felt the house explode as it rumbled through the pavement.

15

# DCI MacGillivray

A cacophony of ravens woke MacGillivray, disturbed by his ringing landline. He'd not gone to his frigid bed upstairs, falling asleep on the sofa after poring over case notes. He'd wrapped his wife's blue throw around him, the scent of her both a salve and a torture. The remains of last night's takeaway brittle in its foil on the table, and a bottle of Black Sheep, hours ago freed of cap and contents.

He hadn't drawn the curtains last night, either. Not that it had mattered; winter darkness reigned behind the window. MacGillivray eyed the clock as he answered the phone: four fifty am.

*It's early.* Though, for the first time in an age, he'd slept deeply.

And his beach-combing brain had sorted through yesterday's treasures, finding things the conscious sometimes miss, picking up on shades and shifts in the clues of the topographical moods of people. He had long ago noticed that humans leave a million traces, and that the subconscious casts its net and hauls them in, and now MacGillivray senses he is tuning in to the detective's engine in his head again.

He picked up his phone and Hanbury gave him the news of the incident at the Leighs.

'I'll be right there.' *Crozier's theory of statistical anomalies showed no signs of abating.*

He ran an electric razor across his face. His reflection in his shaving mirror showed the summation of the years of his life writ large in the grey pallor of his skin. Thankfully, he didn't need to win a beauty competition to attend the aftermath of the Leigh's demise.

CROZIER MET him at the Leigh's

'You look fresher this morning,' she said.

'The after-gig party ended earlier than usual,' said MacGillivray.

She raised her eyebrow in surprised approval. 'Humour, nice.'

He took in the burned-out scene; rivulets of rain ran away, having helped to suppress the fire, now leaving subdued, hissing embers.

'At least this raining cats and dogs weather comes in handy for something,' said Crozier, taking in the scene. 'It's an odd expression: *raining cats and dogs*.'

'A Welsh poet–Henry Vaughn, I think - may have been the first to use the expression. Something to do with a sturdy roof - but don't quote me.'

'If we had the internet back, it would be easy to check.'

'For now, we use the library. Isn't that a nightmare?'

Choosing to press on, Crozier said, 'It looks like a gas leak explosion. Another household that didn't bother to change the batteries on their carbon monoxide sensors. At least, that's how it appears.

There are three dead: Sam, Ally, and the eldest son. James Leigh was out at the time of the incident.'

'Lucky for him. Where is he?'

'In the car,' she nodded to a squad vehicle.

James sat in the backseat. Even through the rain, he looked lost, bewildered. He had curled his body inwards, a natural response to protect himself.

MacGillivray wished he could take the pain away for the lad. *Assuming he had nothing to do with it. And who'd be a police officer, having to think these thoughts?*

After perusing the site, MacGillivray and Crozier left the team to finish, taking James back to the station.

MACGILLIVRAY, Crozier, and James sat in an informal triangle in the interview room at the police station. The boy had sweaty, offish grey skin; shock leaving its impression. He looked fragile, wary, his inner demons no doubt busy chiselling away, fashioning a monument of grief. Something all too familiar to MacGillivray.

'We must go carefully with James until social services can come out,' said Crozier.

James looked up; eyes dilated.

'It's fine,' said James, nodding his assent. 'The sooner I answer your questions, the quicker it's over.'

It isn't the chippy, belligerent tone MacGillivray expected.

'Thanks, James, we can stop anytime you want.'

Mason came in with a blanket, arranging it around the boy's shoulders. He handed him a bowl of broth. James played with the food, gaining a little enthusiasm despite his emotional state.

'Why were you out last night?'

'I'd had a row with the old man.'

'You were out a long time.'

'Yeah, I was angry.' Eye contact disengaged.

*Not a good start,* thought MacGillivray since he wasn't sure he believed him; a glance at Crozier showed she had her doubts, too.

'We know this is painful, but could you tell us what happened when you returned home?'

'I found my home on fire and my family dead.'

He put the bowl aside, pulling the blanket tighter around him, an extra skin and barrier.

The smell of smoke hung about James' clothes.

James looked downwards before saying, 'There's something else I should tell you.'

'Go on, James,' said Crozier.

'Someone tried to hurt me at the warehouse yesterday afternoon.'

'You mean where Bakewell got killed?' MacGillivray struggled to keep the surprise out of his voice.

James nodded. MacGillivray flashed a look at Crozier.

'What happened?'

'A man was going to shoot me. But a window was smashed, which distracted him. Then he fired at the security hut.'

James was back there, the fear of being hunted on his face.

'Then he shot at me as I ran away.'

'Were you hit?' asked Crozier, moving towards James.

James put his hand in his pocket and took out a bullet, its head flattened.

'Yes, only bruising. I found this in my textbook, from my rucksack I'd been wearing. The book had stopped it.'

'Can I take it?' said Crozier. James passed it over to her. 'I'll get it checked over.'

'Did you recognise who it was?' asked MacGillivray.

'No.' He gave a slow, tired shake of the head.

'Can you describe him?'

'Tall, thinnish. He didn't seem too old, not as old as my parents.' He stopped as it hit him afresh; *they were gone.* He looked down, rubbing his face.

MacGillivray gave James a moment to compose himself.

'He wore black clothes.'

'Was there anything unusual about him?' asked MacGillivray.

'No. It was all so quick. Except he knew how to handle a gun.'

*And James would know. He'd probably spent enough time around people with weapons. So, was the killer a professional gunman?*

MacGillivray used his thumb and forefinger to rub harshly at the triangle of flesh at the base of his thumb.

'What time was this?'

'It must've been about mid-day.'

'Where were you when he shot at you, James?' asked Crozier.

'I'd made it down to the river footpath before the bridge when he fired at me. It made me fall off the path. I heard two more shots, but I don't think they were aimed at me.'

MacGillivray tallied the gunfire; The first shot was

at the guard, a second at James, the third at someone on the pedestrian bridge, which had hit the tree, and a fourth finishing the guard.

*Whoever the gunman had fired his third shot at had got away. But who? Perhaps the person who had broken the guard's window to distract him and given James a chance to escape?*

'James, if you can tell us anything else, it might help find your attacker and help us work out what happened to your family.'

James said, 'Me being shot at and my family dying are connected?'

'It looks possible.'

James' story seemed to fit. But MacGillivray still felt that he's holding something back.

He could see James' emotional dam weakening, his jaw working. *Had he pushed him too hard? He'd have to give him a little space. But he needed information, this being the hot time of the case and only likely to get colder by the minute.*

'I never felt I was one of them,' said James, fighting tears. 'I wanted to get away. Because of what they did, who they are. *Were.* It's my fault. I'd wished for them to be gone. But not like this.' Tears broke ranks and ran down his face—salty droplets looking for somewhere to hide.

'You're not to blame.' MacGillivray said, hating himself for considering James could be a suspect. But if he *was* innocent, he needed comfort.

Hanbury knocked and came in on MacGillivray's invitation.

'The parents of Ash Heath have called to report him missing; his bed was not slept in. They contacted his friend Steve Taylor, who thinks he might be at his girlfriend's home at the Richards's place on Orchard View Lane. We rang and spoke to them, and their daughter Angelica confirmed Ash Heath had been there last night. Her parents knew nothing about their daughter's visitor until they asked her. They'd been out then. Heath left around midnight.'

'Another incident. When it rains, it literally pours,' said Crozier.

'We're on our way,' said MacGillivray.

MACGILLIVRAY EASED his BMW through the accumulated water on the road to the Richards' place. 'It's too deep,' said MacGillivray, reversing and parking his BMW.

'He might have cut across the field, past those trees. It would've been a nightmare in last night's conditions,' said MacGillivray.

'Yes, easy to get lost.' Added Crozier.

MacGillivray climbed gingerly over the wire fence into the field, dislodging clusters of rain from the steel barbs.

He made his way up the hill, then around the large copse, cursing the clods of earth sticking to his shoes. Crozier following, uttering less silent curses in Anglo-Dutch, MacGillivray making out a *fucking jammer,* for one.

He stopped to catch his breath, back sweaty and damp. He shouted Ash's name, scanning for any clue where he could be.

MacGillivray spotted a glimmer of incongruous blonde hair.

'Let's try over there,' said MacGillivray, pointing.

'Good shout,' said Crozier.

He dragged his legs as if wading through treacle.

The boy, Ash Heath, is curled up asleep under a large oak tree. He opened his eyes.

'I was dreaming of chocolate. Have you got any?'
asked Ash.

16

# DCI MacGillivray

*Sam Marsh might have a détente with the Leighs, but tough times can destabilise peace. And Marsh could gain significantly from the Leigh's passing, which was why he didn't look surprised to be brought in for questioning.*

MacGillivray is painfully aware that his judgment is coloured by his hunger for retribution: he'd made that mistake of allowing himself to be swayed by his baser feelings before, but he'd failed, trading everything in to do one bad thing, and it had stained his soul.

*Usually, he believed that justice should be sacred, but where had this feeling of justice come from?* He'd burned with the wrong of being blamed and beaten by his father for things he'd not done in his youth, of having to take on responsibilities; like looking after his sick mother, doing all the chores in the house. While his father felt the need to escape to the pub.

That early home life had resulted in the robbery of his childhood–and that had left its mark. His upbringing had provided the bedrock of his reasons for being a police officer, of trying to make things right for others, of feeling the dopamine hit of being the hero.

*Maybe redemption could give him back that feeling again?*

*But this is not a good time to dredge it all up, but then, is there ever?*

He scratched savagely behind his ear.

Sam Marsh had grey bags under his eyes, tombstones commemorating lost nights of sleep and a sallow complexion, a legacy of hitting the bottle during that lost vigil for Tommy. 'Here we go again, Detective Chief Inspector MacGillivray. Trying to pin something on me I didn't do.'

MacGillivray tried to ignore the monster pounding on his inner fortress walls, trying to exact revenge. He breathed, focused, then said, 'You have the motive for killing the Leighs, and that would be to get the town on a platter.' *Especially if the police station gets closed. Who would stop Marsh, then?*

Marsh looked disappointed with MacGillivray's composure but summoned enough bile to say, 'You're not sure what happened to the Leighs. You don't have the evidence to convict me. I can have a lawyer here in no time, and we both know I'll be out. Just let me go and stop wasting our time.'

*One bullet and it would be done, and Sam Marsh could be gone. He gripped the gun in his pocket tight.*

'If it wasn't you, then who could it have been?'

Marsh faltered, tripped up by the question. 'I...I'm not sure.' *He's lying, he does know.*

MacGillivray applied pressure by stepping forward. 'And your son died. You're looking for someone to blame. Who else but Leigh?'

'I hope you can stick to the letter of the Law, DCI MacGillivray. I know you're a better man than me than to fit me up. You've got morals.' His words dripped with sarcasm.

MacGillivray hesitated–*does he know what I did?* And Crozier was staring at MacGillivray. He could almost see her mind calculating, sifting through the sands of what she observed, panhandling for the gold of revelation.

Marsh looked gratified that his comment had hit home. 'It's true. I think my son's death wasn't an accident. And I want someone to pay. But I spoke to Reg Leigh, who convinced me he had nothing to do with it.'

'When did you last see him?'

'We spoke on the phone yesterday.'

MacGillivray scrutinised him, hating that it felt, in his core, that Marsh was telling the truth.

'YOU'VE LET HIM GO. Do you believe him?' asked Crozier.

'Yes, I do. Perhaps the Leigh's deaths are accidents, and just maybe, something else is going on, but I think Marsh knows.'

'I am with you on that.'

'Good, so let's get the team together for a meeting.'

Crozier looked thoughtful.

It bothered MacGillivray that Crozier hadn't queried what Marsh had meant by his comment. *Did she already know what he'd done? Is he being paranoid?*

MacGillivray scratched at his neck, smearing ointment, causing it to sting the abused epidermis. *The body doesn't lie.*

17

# Ash

Ash had strict instructions from his parents to stay at home. *But how could he? There were so many questions to find answers to.*

And he felt fine. *Mostly.* He got out of bed, dressed, and slipped out, heading for Steve's place.

DESPITE EVERYTHING THAT HAD HAPPENED, it astonished Ash that Steve's first question was, 'Was it *worth* it? Going out to Angelica's?' Steve looked admiring, envious even. 'You could have, you know, *died*.'

*Steve believed something was still happening between him and Angelica.* Ash wanted him to keep believing. But he couldn't lie.

'It's over,' he said.

'Oh. Sorry, mate. Did me telling the Police where you were, mess things up?'

'No. It's lucky you did, or who knows what might have happened.' Ash shuddered at the thought.

Steve breathed out, allowing a smile in, looking relieved.

'Why did she end it?'

'I ended it.'

'*You* did?'

'Yes,' said Ash, irritated at his friend's disbelief, 'Come on. We're going to the killer's place, so stop going on about it.'

'Alright, alright, just asking. But... can't you tell me *something*?'

'Shut up!'

THEY WAITED JUST a few terraced houses away from Holroyd and Grey's apartment. Then Grey left the building, heading towards town.

'I don't believe our luck! He's going out,' said Ash.

'What do you mean? I don't feel lucky; I just feel bloody cold and wet.' Ash had to admit Steve had a tinge of blue about him. *It's crazy. Only a few months ago, it had been the hottest summer on record. Now, they were soaked, cold to the bone.*

'Look - there's an open window. Can you be a lookout?' asked Ash.

'Why? What are you going to do?'

'Watch!' Ash went across the road and started climbing the drainpipe. He slipped on the rain-coated surface but held on. Water gurgled next to his ear as a toilet flushed. He felt disgust at what might be inches from his face.

He also felt faint from his fear of heights.

*The peeling wooden ledge looked thinner than it did from below*. He placed a foot on it and tugged at the window. His hand struggling to reach the sash. 'Come on, lift!' he pleaded. His arm ached, gravity using his own body against him.

Then the window budged. Ash, heady with relief, hauled himself into the room. He turned, giving Steve the thumbs up. Despite his nervous exhaustion, he stifled a laugh at Steve's pale face.

*I'm in the killer's apartment!*

He remembered visiting his relatives in Southwold. When his cousin Emily was out, he'd caught a glimpse through her bedroom door. He'd never been in a girl's room. Curiosity got the better of him, or the worst, and he'd gone in. A black bass guitar leaned against an amp, a purple cable linking them. Wide-striped purple and black wallpaper gave off a strong vibe, like the picture of the band on the wall called Blood Red Ellipse. The vampire poster art and goth jewellery helped to make the room cool. It was strange - the longer he'd been in there, the more uncomfortable he'd felt, his fear of being caught escalating. He knew he would have hated it if someone had done the same to him, gone into *his* room, gone through *his* stuff. Not that he had done that - he drew the line at least there. He was just *curious*. But later, he'd been so *disappointed* with himself.

*Here, in the killer's place, it's different. If he got caught, disappointment in himself would be the least of his problems.*

One room had an unmade bed, white sheets, and a black duvet. A few dark clothes are hanging in the wardrobe. In the second bedroom, a blue and white box of tampons sat on the bedside table; he knew what they were from having to go to the chemist to get them for his mother, much to the sniggering mirth of some girls from school. 'Heavy month?' one of them had cajoled, causing the others to burst out laughing.

He walked across the carpeted floor, opened the door, and entered an open-plan lounge and kitchen. A faint smell of a stew came from a pot on the kitchen hob.

In the lounge, he saw a beige folder on a glass coffee table. He sat on a light green leather sofa, perched on its edge, and opened the file. It revealed a list of names, with lines struck through some of them. He recognised some people who had died.

*Is it a kill list?*

*Is this the lair of serial killers?*

He rummaged for clues about who they are—finding little joy.

The longer he stayed, the more nerve-jinglingly scared of being caught he became.

*It was time to go; he could not stand the tension anymore. At least it's easier leaving this time through the front door—the prize of the papers in his hand.*

---

NOT LONG AFTER, Ash and Steve sat in the Cosy Igloo café.

They sipped on sweet hot chocolate. Ash felt its calming effect on the hand-shaking surge of adrenaline from breaking into the killers' apartment.

'Some deaths in the town, they're no accidents - we've got proof, look,' said Ash, showing him the papers.

'And there,' Ash pointed towards James Leigh's name. 'It's got the date and time of this afternoon. That's when they're going to get him.'

'Don't tell me - we're going down there.'

'Yep.'

'Do you know the name of the killer?' asked Steve.

'I couldn't find anything on them,'

'Them?'

'Sorry, yes, there is someone else living there. It might be a woman.'

'Two of them? A team of serial killers? Bloody hell.'

*Could it be Miss Holroyd? Was that so crazy?* Thought Ash.

'What if it is the Police? No one goes to the Police. No one trusts them. My dad says they are corrupt,' said Steve.

'This could prove it,' said Ash.

Ash saw the last page, eyes widening.

'What is it?' asked Steve. Ash turned the page around and pointed to the bottom. Ash's name was scrawled in ink—an untidy addition to the typed print.

'Bloody hell, Ash, how does it feel being a marked man?'

'Thanks, Steve, that's just what I needed to be asked. There's something else I've got to tell you. I've seen Miss Holroyd before.'

Steve looked thrown. 'You know, the temp maths teacher.'

'In your dreams, I reckon.'

'No, it's true, she was at my last school.'

'You kept that to yourself?' Steve snapped at him. Ash wished he'd told him sooner, but he'd needed time to think about it.

'I can't swear to it, but I think I saw Miss Holroyd push Harry Butler down the stairs, too.'

'A few days ago, I would've said you imagined it. But now, I'm not so sure. Could she be the one living with the killer?'

'I thought of that, too. It's fantastic, isn't it? But it fits.'

'It's all getting too big,' said Steve, eyes getting bigger.

'Yeah. We should tell someone.' *Could he tell the police officer, MacGillivray? He seemed alright - for the police. And he had found Ash out in the field near Angelica's.*

*Angelica–he hadn't thought about her in a while.*

'I am not sure about that idea. You know you can't talk to the police; you'll be a grass,' said Steve.

Ash's dad talked about *being between a rock and a hard place. Now, he thought he understood what that meant.*

*But he knew he had to do something.*

---

ASH WOULD HAVE LIKED to have had Steve with him for support, but Steve had made it clear that he was dead against what Ash wanted to do. *So he's on his own.*

The Police station is a Victorian red brick building, its metal windows obscured by black wrought iron bars. It looked ominous and miserable, and he felt increasingly exposed as he climbed the worn steps to the front door, rivulets of water cascading down them.

The wind didn't help, blowing the rain into his eyes, forcing him to bring his hood down tighter.

*Turn back. No one will know,* said the little voice inside of him.

18

## DCI MacGillivray

Simon Alderson had been MacGillivray's predecessor, who'd held the rank of regional coordinator. In MacGillivray's opinion, he'd been a good man. *Good to him, anyway, always having time to mentor him and trusting his judgement.*

After the deaths of Yvette and Lucia MacGillivray, Alderson had sent MacGillivray home on compassionate leave. With nothing but death and loneliness to mull over, it hadn't seemed compassionate to him.

After a week of climbing the walls at home, MacGillivray convinced Alderson to reinstate him. The team had been running on anaemic staffing levels even before he went home, so it wasn't such a hard sell. But Alderson shouldn't have allowed him to return, but he had.

Then Alderson died of a heart attack. Maybe the force had taken too much from him.

Alderson's workload, he was told, was now his. Though not the rank - he would stay with his three bath stars as chief inspector - no argument – not that MacGillivray could be bothered to make one.

Soon enough, as Alderson must have experienced, pressure came from on high to play the numbers game. to close cases, to meet the metrics. *It's always about the metrics.* He could smell the fear of his scrambling superiors as they fought for the diminishing places in the political lifeboats as more heads rolled.

Then there's the trialing of robots in policing in the capital. The Government and the NSSA were keen to roll out new tech and AI; the future of policing as they saw and sold it. MacGillivray didn't think the technology was advanced or as bug-free as they implied. The launch of soft and hardware hardly ever were–*wasn't it typical for there to be an average failure rate of 70%?*

But the signs had been encouraging, with some quelling of activities achieved with entry-level models that act as barriers to protect lone officers, and reports of some bonds being built between flesh and metal police officers.

But understandably, away from the inner cities, it added to his team's insecurities and affected their already battered morale. They saw the threat of a human police presence being phased out in favour of dutiful, unimpeachable, 7-days-a-week lawkeepers.

However, he'd been told that the provinces were safe in the medium term because the cities came first, so he needed to sell that idea to his team, and fast.

BUT MACGILLIVRAY'S re-awoken intuition couldn't ignore the idea that he'd been promoted because he was the weak link. Deep down, he knows he didn't deserve to get the step up. Maybe in the past, he'd been *on it* enough, but even he knew he wasn't performing at his previous higher levels. Crozier should have got the promotion. *How bitter must she be? Yet, she still looked out for him...*

*So, was he promoted because he was seen as manipulatable?*

MACGILLIVRAY ADDRESSED HIS ASSEMBLED TEAM. 'We've had an updated forensics report on Tommy Marsh and Eddie Bridges; they have since found bruising in keeping with someone gripping their arms. The size of the marks shows they were large hands. Possibly male, tall, considering the angle of the bruising. Low temperatures of the river water had delayed the onset of ecchymosis.' *People pressured into rushing investigations make mistakes.* 'I now consider their deaths highly suspicious, so we're reopening the cases,' said MacGillivray, raising a few eyebrows.

'On to the Leighs' incident. The explosion killed most of the family, all except James Leigh. He had been fortunate enough to be out at the time of the incident. The gas company is running checks and is obviously keen to get to the root cause.

But, while the gas explosion as the primary cause looks plausible, we must keep an open mind at this stage. Interestingly, there's been a killing not a half mile from the Leighs. A man had his arm broken and his throat cut. He'd been on the edge of Folly Park, though it looked like he'd been dragged there. It's not an unusual outcome for someone out on their own after dark. But it happened about the same time as the Leigh incident. The deceased's name is Anthony Parker, who has a history of rape and mugging.'

'It looks like he was a perp who had picked on the wrong person to victimise,' said Crozier, with a thin line of a mouth that battened down a smile.

'You think Parker was killed by someone who'd

murdered the Leighs? That's a stretch, sir, isn't it?' asked Grey. But Grey looked shaken. *Why be so affected by the idea?*

'Does this mean we are discounting a feud between the Marshs and the Leighs?' asked Hanbury.

'Good question, Hanbury. Nothing is to be ruled out, and be wary of being seduced by naïve assumptions.'

MacGillivray's hypocrisy made him squirm, considering his recent cut-and-paste conclusions about the security guard Bakewell and - heaven help him - maybe other cases, too. 'Keep your eyes and ears open for signs of anything odd, maybe loose talk about another gang moving into town. If there's another influence at play, don't play the bloody heroes and go it alone, is that clear?'

'Yes, sir,' echoed many voices around the room.

MacGillivray relayed James Leigh's version of the Bakewell shooting at the warehouse. Perturbation graduating into astonishment across most of the team. Again, Grey's reaction lacked the surprise that everyone else exhibited. He filed that observation away.

'A gunman shot James Leigh, so it's not a warehouse raid. So, was Keith Bakewell unlucky to be caught up in it?' asked Hanbury.

'It's a possibility,' said MacGillivray. 'DI Crozier and I met with Keith Bakewell's wife today. We think it's premature to conclude that Bakewell is complicit.' *Going by a few guilty looks, he wasn't alone in thinking that.*

'Ask yourself, are we missing anything? Have we turned over enough stones?'

Grey, his voice a cross between a whine and a

snipe. 'Didn't you imply there's nothing to follow up, that it's a waste of budget?'

Jacobs nodded in agreement. Crozier looked sideways at MacGillivray while the rest of the room looked downwards at what felt like a challenge to his leadership.

'The killers could get away. I need people who'll step up. If anyone disagrees, then come and discuss it with me.' The team looked stunned; this was not the usual laissez-faire approach they had grown accustomed to from a PTSD-affected MacGillivray. MacGillivray gave them a moment to process, scoping his depleted staff and looking for signs that he had the room.

The silence stretched, and then Grey said, 'Sorry, sir,' though his arms were tightly folded.

MacGillivray nodded at Grey, then continued. *Maybe a battle had been won*. 'Crozier has information that might shed light on the death of Bakewell.'

He looked towards Crozier. She said, 'Forensics confirmed that the bullet we found in the tree at the warehouse matches the two retrieved from Bakewell's body and James's textbook.'

'All from one gun? So, one shooter, then,' said Hanbury.

'It appears that way. I want the associates working for the Leighs interviewed. Mason, can you and Hanbury oversee that? Let's put the pressure on and see if anyone cracks,' said MacGillivray.

'Sir,' said Mason, straightening.

'Jacobs, anything on the death of the Butler boy at the school?'

'We have struggled to locate Miss Holroyd, as she has not been home.'

*Knowing Jacobs, he'd probably only tried once. Jacobs'*

*minimalist efforts were becoming a growing problem.* And MacGillivray had his suspicions that Jacobs was the leak in the office.

'Don't close the case yet. I want you to take another look at his father's car accident while you are at it.'

'DI Crozier will partner with you.'

MacGillivray caught her eye, signalling with a nod he wanted an eye kept on Jacobs. Crozier gave a curt nod back.

'Any other business before we move on?' asked MacGillivray.

'Eh, yes,' said Jacobs. 'Will machines be taking over our jobs in the field?'

'You mean the Peel Bots,' said MacGillivray.

'After Robert Peel? Isn't he said to be the father of modern policing?' asked Mason.

'That's right. I'm glad someone knows their police history.'

'Smart arse,' said Jacobs.

MacGillivray continued, 'The Bots are only being trialled in the cities now to help control the curfew. It'll be a while before AI replaces human intuition countrywide, though. In the meantime, we have enough to keep us humans busy. So, let's get on.' He shot Jacobs a look, dangling the carrot for him to say something else. Jacobs stayed quiet. *He looks ill, the way you do when you get bad news.*

MacGillivray could feel the old furnace burning within him, feeding on his self-doubt, grief, and apathy. *It's him, the old MacGillivray, good to see.* MacGillivray's problems had allowed Jacobs to drift under the radar for too long. *That had to change.*

'Jacobs, could I have a word in my office, please?'

. . .

JACOBS HAD BEEN *a decent enough police officer. And sometimes, he still did a job. But, since the cutbacks and the increasing threat of Peel Bots, he had become less effective.*

*And, when talking to Jacobs, he had a way of looking through you as if reading a teleprompt.*

'There's a leak in the team, Jacobs.'

'Who have you got in mind, sir?' asked Jacobs, stiffening, his eyelids increasing their blinking rate.

'Have you been passing information to anyone outside of the office?'

Jacobs looked affronted, and the directness of MacGillivray's questioning seemed to have thrown him out.

'You mean *me*?' Jacobs' jaw tensed. 'Eh, why should I give away information?'

He nodded while he talked, as if the act of nodding would persuade MacGillivray to accept what he said. *MacGillivray was not the only one who knew basic influencing skills.*

'You've not answered the question. Let me make it a little easier for you. Have you ever tipped off a suspect that they'd be raided?'

'No, sir.' again, the glazing, tranced look. *A conscious effort to keep looking forward. There's something even creepy about it, a sign that he's trying to hide something.*

'That is a serious accusation, sir. Have you any proof?'

'This is an informal chat. It's your chance to tell me before it's too late.'

Just for a moment, MacGillivray thought he'd own up. But now he saw a man with few options. And confessing wasn't one of them.

'I take it you don't. Have proof, that is, sir. If you have nothing else to ask me, do you mind if I leave?' asked Jacobs, looking peeved. Or worried. MacGillivray couldn't quite tell.

'This offer of leniency will not last. Don't waste the opportunity. A storm's coming, and I need you with us to face it.'

Jacobs sneered, 'The storm's been here for a while, sir. Only you haven't been paying attention.' He took quick strides to the door, gone like a deer running away from danger.

AFTER JACOBS LEFT, the phone rang on his desk, the reinstalled landlines cluttering the world with cables, mobile technology forced to the back of the evolutionary queue.

'DCI MacGillivray,' he said.

'Pendle. Head of National Security Surveillance and Action.'

*Pendle? The NSSA? What did he want?*

Crozier had nicknamed the NSSA, the SS, after the Schutzstaffel; the paramilitary organisation headed by Hitler. Not without reason, and she wasn't the only one wary of them. The NSSA is a shadowy wing of the police security forces, charged with a brief to clean up *The Force.* The NSSA had been responsible for turfing out corrupt officers. And they reported a success rate MacGillivray considered too good to be true. It looked increasingly like an engineered excuse to eliminate those not fitting in with the culture. And rumours circulated of the NSSA's clandestine, hard-line activities, dealing with criminal and dissident factions. It was all refuted by the Government, defending the

NSSA, claiming the media outpourings as malicious left-wing pot-stirring.

Not that it mattered, as that flavour of media had been conveniently closed because of crippled comms and the near-impossible task of distributing print in a failing national infrastructure. Caused by, some thought, the NSSA Argentinian Junta style *Dirty War* tactics. Hard to believe for many, but the rumours persist.

'I'll get straight to the point, MacGillivray. We have heard good things about you. We want you to join us.'

'What does the role entail?' MacGillivray struggled to keep the cynicism out of his voice, scratching the back of his head, dry skin falling like snow.

'Besides existing duties, you'll carry out covert tasks from time to time.'

*What does that mean?* 'Could you elaborate on that?'

'Shall we say you'll be required to handle sensitive matters that must be resolved with delicate handling? And you'll have access to additional resources for the *Marsh* case.' He paused pointedly. 'To complete what you...eh... tried to get across the line.'

*Does he know what I did? Marsh, Crozier and Pendle...does anyone not know what he had done?*

Shortly after Yvette and Lucia's deaths, he'd gone around to the Marsh's place on the pretext of preliminary enquiries under the auspices of a warrant. While there, he grabbed one of Marsh's hats hanging in their hall. The same hat Sam Marsh had been wearing from the CCTV footage.

He'd visited the morgue and talked the technician into going against protocol, giving him a few moments alone with his wife and daughter. They both lay on neighbouring steel pull-out shelving. *Even in death, their fingers seemed to reach across the distance between them.*

Then he tainted Marsh's hat, smearing his daughter's blood and attaching her hair.

And on another contrived visit to the Marsh home, MacGillivray had gone to their garden shed, *and, forgive him, he had planted the evidence. Because without it, there wasn't enough proof to tie Marsh to the murders. He'd traded everything for the certainty of sending Marsh down. And wasn't MacGillivray on the side of the angels?*

But later, before the court case, the hat had mysteriously disappeared from the evidence storage, along with the DNA tests—*surely too many coincidences for this to be incompetence?* The word *betrayal* sounded too harsh, too brutal. But that conclusion felt inevitable.

Whatever the truth, losing evidence had been enough to give Marsh his escape from justice. And If Pendle could prove MacGillivray had planted it, he'd be finished. *He needed Pendle's silence.*

'If you reach required performance levels, closing specific cases promptly, your station will enjoy a healthy budget.'

*And if I say no and the station shuts, what'll happen to the town? At the mercy of rising crime, dependent on an overworked Crawley police force, who are not so much a force as a trickle.*

'Without labouring the point, you will need to keep your role with the NSSA *discreet*.'

*He could save the police station, save jobs, and have the budget to continue to chase Marsh. And save himself from disgrace.*

*How could he say no? But it's a deal with the devil.*

'WHAT DID HE SAY?' asked Crozier.

'To give me a pep talk.' *It continues: one immoral act leading down to another - more lies.*

'The new fresh face of police management, right?' Crozier said though she didn't look convinced by his reply, a pinch of a frown casting a small valley of shadow on her forehead. *She isn't a DI for nothing.* 'I am calling on Miss Holroyd.'

'While I'm lumbered with Jacobs,' said Crozier.

'You'll be leading your own team soon enough. You might as well get used to managing difficult team members.'

Crozier could not disguise a smile, or her ambition.

MacGillivray grabbed his coat to leave. Then, through his office window, he recognised a schoolboy furtively approaching the station. *He looked familiar. The boy from the school–it's Ash Heath. It must be important for him to come to the police station of his own volition.*

Then he remembered he'd wanted to see Ash Heath to ask if he knew anything about Tommy and Eddie's deaths. *But that's now a redundant line of inquiry.*

*So why is Ash Heath here?*

HE MET the boy at the entrance door, opening it for him to come in, rainwater dripping from him onto the tiled grey floor.

'Hello Ash, what brings you here?'

The boy looked pale, fledgling anxiety lines vying for permanency on his face.

'I want to thank you for finding me in the field,' said Ash.

'You're welcome,' he said with a smile. 'I'm surprised to see you up and about so soon. You must have been close to catching your death out there.'

'I'm a fast healer,' he said quickly. 'I've got things to tell you, but I'm unsure where to start.'

'You've watched a few cop programmes, they would say...'

'Start at the beginning?' MacGillivray nodded.

'Let's go to my office. Let me get you a drink?'

THE BOY, nursing the drink, sat facing MacGillivray in front of his desk. The boy seemed to muster his thoughts, then plunged in. 'There's a teacher at the school called Miss Holroyd.'

*The elusive Miss Holroyd.*

'I've seen her before.'

'In the town?'

'No, I mean, at my last school in Warburn. Before we moved *here*.' The lad's contempt for Hawham piled thick.

'Okay,' MacGillivray said, that old excitement building at the potential for connections. 'Go on.'

'A boy died falling down the stairs at that school. And Miss Holroyd had found him there, too.'

*Miss Holroyd?*

'That's probably a coincidence, Ash.'

'Yeah, well, I think I saw her push Tommy down the staircase.'

MacGillivray stared at him, taking in this new piece of information. 'Are you sure about what you saw?'

'Yes, I think so. And that's not all,' said Ash, words

quickening, 'I saw the man who killed Tommy Marsh.'

'How can you know?'

'I saw it happen.'

He recoiled in his chair—*bloody hell.*

'Tell me what you saw.'

Ash described the events at the river.

*As a working theory, it's reasonable to assume that the mystery person had killed Eddie Bridges, too.*

'That's not all. I saw the killer in town later and followed him to his home.'

'What's the address?' asked MacGillivray, lifting a notepad and pen from his desk drawer.

Ash told him, adding about the signs of someone else living there. MacGillivray raised his eyebrows. *Wait a moment; he'd seen the address before.* He scooped up Holroyd's file. *Yes, here it is; it's Holroyd's address, too. Holroyd inhabited the same apartment as the killer. So, what was her role in all of this? And was Harry Butler's death really an accident? And therefore, what of Harry Butler's father?*

'How could you know someone else lives there?'

Ash took short, quick breaths, eyes widening, nervous at this confessional.

'I climbed into the killer's apartment.'

'You did *what*?' Ash leaned back in his chair, taken aback by MacGillivray's reaction.

MacGillivray gathered himself. *It would do no good to spook the boy further than he already is.*

'Sorry, Ash, relax and tell me what you know.'

Ash placed a file on the desk. MacGillivray picked them up and read through them: a file of names, surnames, struck through of those that he recognised had been killed. *Deaths he had rushed to close as accidental.*

'Am I in trouble?'

*MacGillivray isn't about to give Ash the notion that burglary is acceptable.* Even though, despite himself, he's impressed. *But what if they'd come back and caught him?* He shuddered at the possibility. He shuddered, too, at what this list meant.

'You did these things for the right reasons.'

He saw the boy's worried frown ease a little.

'But we will still need to have a chat when this is all over.'

Ash nodded, understanding.

'Can you describe the man?' asked MacGillivray.

'I have these.'

Ash pulled his old phone from his bag and showed MacGillivray his pictures.

Even with the black hat covering much of his head and face, the man looked annoyingly familiar to MacGillivray.

'Do you know who he is?' asked Ash.

'Not yet, but we will do.'

'What do we do next?' asked Ash.

'*We* are not doing anything. *You* are going home to stay out of trouble.'

Crozier appeared at the glass panel of the door. MacGillivray signalled for her to come in.

She smiled at Ash, then said to MacGillivray. 'Jacobs has gone home. He told Mason he wasn't feeling well. Whatever did you say to him?'

# 19

# Holroyd

Holroyd sat in the lunar glow of the nightlight that turned the rich colour of the sofa to a sun-starved winter green as she waited for Grey to return.

She heard the key in the door, followed by Grey's silhouette, backlit by the landing lights, as he entered the apartment. Grey flicked the light switch, illuminating the open-plan kitchen, lounge and diner of their apartment.

Holroyd sat on the sofa next to the empty file, her gun pointed at him.

Holroyd could almost see Grey's brain processing this new information.

'Ah, I forgot to lock the files away. The empty file suggests they are missing,' said Grey.

She nodded.

'And you had no choice but to tell Pendle. And now you have orders to retire me.'

She nodded almost imperceptibly. *She didn't want to kill him. Only he'd become such a liability. And he'd take her down, too. But perhaps if she'd waited, she might have calmed down from her exasperation and done something different. But she hadn't.*

*And, usually, she'd have pulled the trigger at a confirmed target. So, what was she waiting for?*

'But here we are.' Grey's eyebrows raised in surprise at not being dead.

*Because their relationship had changed from the clinical to a cluttered something.*

'I've considered a life with you, leaving this behind. But that's all a bit late. Save us from clichés, right?' said Grey.

*There, it had been said, now out there.* Her softening, pained eyes acknowledged the idea had crossed her mind, too.

'But we're in too deep. We would spend the rest of our days running,' Grey added.

Again, she nodded, hating that she'd been burdened with these feelings of sadness.

Their exchange, the last item on the agenda of Grey's life.

But the moment held, suspended. *Why the delay? Pull the bloody trigger,* she implored herself. She kept giving him second chances. *Damn you, Grey.*

Grey had kept his hand on the light switch, and now he plunged the room back into darkness with a simple finger movement.

Holroyd's bullet found the wall behind where he had been. Grey might have beaten lightning to the hall. The flash of light from her gun blinded Holroyd in the darkness. *Damn.*

Holroyd heard two rapid shots fired, accompanied by the shattering of the landing window.

She got to the hall and saw Grey follow his bullets through the smashed opening.

She ran and looked out of the room window. He'd landed on a raised saturated grass bed and got a shot away at her, causing her to duck back. She risked a peek through the opening. Grey had scrambled to get a vehicle engine between them, one of the few things that could stop a bullet.

Holroyd pulled the trigger. Its hammer smashed on the bullet, causing an explosion, propelling the metal from the chamber. She felt the familiar recoil through the gun. The panel of the car Grey hid behind buckled, complained as her bullet hit. Grey shot back, causing her to retrace a step.

*It would be suicide to follow him out of the window, to be so exposed.* Instead, she opted to head downstairs to the main entrance. *She had to stop him.*

## 20

## DCI MacGillivray

A charm bracelet swung sideways from the rear-view mirror, a silvery horse with a magic horn protruding from its head. It was a home-made birthday present for MacGillivray from his daughter, and she'd written in her card that '*Unicorn would look after daddy.*' The bracelet had constantly reminded him to be as safe as possible for his wife and daughter's sake. Now, it only reminded him to do it for his own, prompting him to consider if it's sensible to be here solo, to contradict the instructions he had given his team. He promised himself he would only reconnoitre the site.

Holroyd's apartment block hove into view.

Then, the sound of gunfire. The downpour had stopped, allowing the sound to echo in the rain-scrubbed evening.

Adrenaline kicked in and he pulled his gun from his glove compartment.

There was a flash of movement outside; someone yanked open the passenger side door and jumped in —with the smell of stale cigarettes.

'Grey? What are you doing here?'

Then he saw the profile of Grey with his hat, wondering why it fitted a question that had been loitering in his subconscious. Then the penny dropped, Ash Heath's photographs, *Grey is the mystery man in them, why he looked so familiar.*

*Had Grey killed the Marsh and Bridges boys*? MacGillivray's brain crunches with the gear change, and he turns his gun on Grey.

'Not a good move, sir,' the gun already pointed at MacGillivray. MacGillivray rested the weapon on the dash slowly, putting his arms up in the customary *I surrender* position, his fingers touching the cold surface of the roof.

A familiar blonde woman came towards them, a gun an extension of her arm. *Miss Holroyd.*

'What is going on?' He asked again, trying to exert authority over Grey. But suspected that ship had well and truly sailed.

'Drive,' said Grey quietly, pointing his gun in MacGillivray's face. 'Fucking drive, *now*.' MacGillivray heard the click of the weapon; it was claustrophobic in the car now.

A bullet rattled the bonnet of MacGillivray's car.

Holroyd is wearing a snarl of aggression and concentration as she raises and fires her handgun, another bullet whining off the car's chassis.

Suitably compelled, MacGillivray struck up the engine and reversed along the street, his anxiety causing the vehicle to slide on the wet road.

Holroyd fired at them again; MacGillivray's left-wing mirror disappeared.

He made it into a side street and jammed the car into a complaining forward gear before it stopped rolling back. His stubborn preference for old manual change cars threatens to be his undoing. He gunned the accelerator, causing wheels to spin as the tyres fought for traction on the slick surface.

'For fuck's sake, take it *carefully*,' said his passenger, transforming from his diffident persona to something more treacherous.

MacGillivray saw their attacker in his rear-view mirror, stopping, adopting a shooting stance. She fired shots in quick succession. The rear windscreen shattered, glass scattered, wet air and the pronounced sound of MacGillivray's overworked engine flooding the interior.

He aquaplaned the car up the long street over a sheet of never-ending flood water, approaching the junction too fast.

'Right?' He posed, instinctively wanting to home in on the police station. *But Grey might have other plans.*

'Yes, and slow down!'

MacGillivray took the junction too quickly, a triumph of fear over ability. The car's wheel found a deep cavity of broken tarmac hidden in the floodwater. The BMW swerved, adhesion lost, colliding with a parked yellow Subaru, buckling wing panels and locking wheel arches.

Grey's head had slammed against his side window. He snapped back, dazed. Blood oozed from a cut to the side of his forehead. His gun hung loosely from his right hand. And MacGillivray noticed dark blood flowing from his right shoulder, having sustained what looked like a bullet hit.

MacGillivray, without seatbelt on, had lurched at an angle, crunching into the dashboard.

He tried to pry the gun from Grey's hand, but Grey's fingers clamped on it, even in a semi-conscious state.

MacGillivray tried to start the engine. *Nothing. The car was going nowhere.*

MacGillivray gingerly got out, auditing himself in a micro-moment for injury, declaring himself fine.

He peered down Lion Street, spotting Holroyd in the distance running, with an elegant gait towards him, gun in hand. Bullets engraved with their names.

He couldn't dislodge Grey from the passenger door as it had mashed against the parked vehicle, and Grey is too heavy to pull out via the driver's side. MacGillivray reached for his gun only to find Grey pointing his gun into his face, waving him back. Grey slid across the driver's seat, navigating past the gear stick. He stood, cradling his gun, his jaw clenched with the pain from his gunshot wound.

'I said slow down,' he said groggily, left hand touching his temple, finding blood on his fingers.

'Sorry,' *why was he apologising?* 'We must get going. Holroyd is coming for us, but the car's dead,' said MacGillivray, looking back. Lucidity returned to Grey's eyes, forcing himself across the seats to get out of the driver's side. Grey stood up and shook himself like a dog. *He's one tough character.*

'Let's use my car; we can take a back route to it,' said Grey.

*It's a good plan, and hopefully, we'll lose Holroyd for long enough.*

'Good idea.'

*Why did Grey want him with him? Why not leave him for dead? Or even shoot him? He'd find out soon enough.*

Grey led the way. Stopping at a bus shelter, demanding that MacGillivray rip off part of his shirt. MacGillivray did as he was told and gave the strip of material to Grey. He held the cloth against his gun wound and applied pressure to stem the leak of blood. Then Grey staggered on, signalling with his gun for MacGillivray to follow.

## 21

# Holroyd

Holroyd arrived at the junction. Breathing hard, gun arm ninety degrees to her body, swinging from left to right, scanning for her quarry. She is wary of crossing the exposed road, noting the stillness, both an ally and a threat. She edges along a line of parked cars before taking the gamble to cross. Holroyd closes in on the crashed vehicle. The rain is now a cascading symphony of drops bouncing on metal roofs.

*Still no retaliation, no ambush.* She continued her cautious approach, acutely aware of time leaking away. She welcomed this adrenaline, the thrill of being *out there*–she just wished it wasn't Grey she had to chase. *And what about the police officer? Had it come down to killing detectives, now? This was one messed up situation.*

A hiss offered itself, a fractured radiator leaking steam from the crashed motor.

*The police officer's car is empty.*

*Fuck. Where would Grey and the detective go?*

Holroyd looked up the road, trying to peer through a curtain of falling water.

She couldn't see them. She headed towards the

town centre, seeking signs of where they might have gone.

*There. She almost missed them, splashes of blood against a bus shelter, little splashbacks of rain diluting its hastening run down the glass. They'd taken a side street. They had doubled back through the back roads...to... Grey's car, of course.*

She took off, electric urgency coursing through her.

# 22

# DCI MacGillivray

MacGillivray and Grey had returned onto Lion Street, across from Grey and Holroyd's apartment.

MacGillivray's instincts had been right; Grey had not been what he seemed. *If he's not emergency help from a police pool, then what is he? And if he's not the police, then how did he fool the selectors? Is he a serial killer? Is he working for a criminal gang? Or could he be working for an agency? And what of his connection with Holroyd, the assassin teacher? And why are they trying to kill each other now?*

'The blue Land Rover Defender 3,' said Grey, nodding towards the vehicle parked on the grounds of the apartment block.

MacGillivray could make out Holroyd, making her strident way from the junction. She hadn't been fooled for long. 'She's coming,' he said.

'I've lost the frigging key,' Grey's cool, warming, as he rummaged through his pockets. 'No, wait, here they are.' Pointed and clicked, door locks popping.

He threw the key at MacGillivray. 'You're driving.'

*Grey is badly injured and in no fit state to drive - so that is his purpose for Grey. At least he would live long enough to drive Grey to his destination. Since his family's deaths, he had barely wanted to exist, but now he clung to every moment like they were precious jewels.*

A few residents risked their heads above the parapets of their castles, curtains flapping. At the sight of the bloodied Grey, they lost interest quickly, fearfully retreating.

MacGillivray could hear a distant, emphatic police siren, willing to serve, promising to save. Its sound muffled by buildings.

Their pursuer stopped about six car lengths away. Holroyd swung her gun arm up.

MacGillivray pressed the starter button, slamming the automatic into D and pulled out.

Holroyd fired. The Defender's rear window succumbing, causing Grey and MacGillivray to duck.

MacGillivray managed, this time, to finesse the car without hitting parked vehicles. In the rear-view mirror the blond woman kicked a car's tyre, a study in frustration.

## 23

# Holroyd

Holroyd is cursing, hands on her thighs, sucking in wet air into sore lungs, her hair hangs dank with sweat and rain about her face as she watches Grey and the police officer escape, the air pungent with Grey's car fumes, seemingly mocking her.

WITHIN MOMENTS, she stood, feet cold, in a piss-addled telephone box. Holroyd caught her reflection in the uncracked mirror; *perhaps vandals could be narcissists, too.*

Pendle answered her call straightaway.

'Holroyd,' she said.

'What is it?'

She gave her tidings, unsweetened.

'It's a fucking nightmare,' he said. 'I'll call you back.'

She didn't have long to wait.

Holroyd punched the receive button on the telescreen. A nanosecond after the call came through. The video option, as usual, ignored by Pendle. Not that she wanted to see his face anointed with his fury; she could hear that in his voice.

'We're sending a team in to cleanse the site. You are to find Grey and deal with him. Be in no doubt - your future is at stake.'

'What about the apartment?'

'We're compromised. Set the building alight. Leave no trace.'

*Call ended*, announced the screen.

*The police siren's getting louder, nearer.* Holroyd didn't have long.

She retraced her steps to their apartment and set fire to the curtains, hitting the glass of the fire alarm panel on the way out while hearing the agitated voices of residents responding to the baying alarm.

Once outside, she looked back. White and yellow tongues of flame move with alarming speed, causing windows to shatter, allowing oxygen to rush in, an eager ally to fire - that indiscriminate, egalitarian force, practising its bewitching transition of materials from solid state to energy.

Voices shrill with escalating panic urged escape, to leave everything behind, hastily dressed people in slippers, hastily negotiating the external black steel fire escape, gathering in frightened clusters outside their burning homes.

In time, the survivors would harvest the stings of tragic deficit as their inventory of loss grew; photos, artefacts, favourite clothes, all lost, mercilessly burnt, in the arc of the fire's pyrotechnic life.

She slipped into her silver Volkswagen Polo. The luxury, sealed-from-the-outside-world metal box earthed her, feeling her head clear with each breath, sorting the wheat from the chaff of her renegade thoughts.

*Where would she find Grey?*

## 24

## DCI MacGillivray

The man in the sandy-coloured corduroy trousers and green and silver Fleur de Lys waistcoat turned to MacGillivray and said, 'Well, detective, the bullet has gone straight through his right shoulder, from which he should recover, but he has sustained a concussion.'

'When can we question him, Doctor?'

'Mr Hughes,' he corrected, surrounded by an ensemble of stiff-backed, terrified staff. *This man had the power to end their careers - as if the NHS could afford to lose more staff to the private sector.*

'He is to rest. Let us see how he is when he comes to,' he said, a guarded concession offered.

'Thank you, Mr Hughes.'

MacGillivray stepped out of Grey's private room into the path of a rushing Crozier.

'What's going on?' she asked.

MacGillivray told Hanbury, who stood guard, 'We're going for a coffee. Let us know if Grey awakes.'

THE BEVERAGE VENDING machine dispensed its brown, burnt-smelling liquid that masquerades as coffee, but he still cradled the cup, revelling in its emanating heat.

He relayed an account of his visit to Holroyd and subsequent events.

'You've had quite the adventure. So, PC Jason Grey isn't who he appears to be. You were never sure about him, were you?'

He raised his eyebrows, surprised that Holroyd had picked up on MacGillivray's suspicions. 'It doesn't do much for morale if I cast aspersions about the team, does it, DI Crozier?'

'Perhaps not,' she conceded. 'And Holroyd, the teacher from the Forêt School, she's involved with him. So, she might have killed the boy Butler and his father, too. *Are* they serial killers?'

'I don't think so, at least not how you mean.'

'What do you mean?'

'I have not worked it out yet. However, I think Grey killed the two boys by the river, possibly to spark a war between Marsh and Leigh. But I am not sure who Grey represents in this scenario.'

'Why didn't Grey kill them himself? Wait a minute, I see. He'd get the two families doing the dirty work.'

'It looks that way.'

'Who do you think is this third party? Another gang muscling in?'

'Perhaps,' But MacGillivray's suspicions were becoming wilder, ones he isn't ready to share, not even with Crozier.

Hanbury interrupted them, 'There's an urgent message for you from Pendle, the Head of the NSSA. He wants you to meet him at the entrance hall now, sir.'

*What a coincidence.* Thought MacGillivray.

'It's got to be a big deal. If the head of NSSA is calling.' said Crozier as they took the lift down. MacGillivray didn't need telling.

*He should lay out everything they'd found with Pendle. But...*

---

TWO BLACK GLASSED LAND ROVERS, all chrome fittings and pomposity, had stopped under the hospital entrance canopy.

'The NSSA were clearly unaffected by any budget restrictions, a stark contrast to the run-down hospital, don't you think?' said Crozier, echoing MacGillivray's thoughts, but there is no time to comment.

Two of Pendle's team got out of the second Rover, wearing dark blue uniforms, all sculptured muscle and focused, darting eyes. Moments later, Pendle followed. *He's arrogance in a uniform, six feet two inches tall, dominating adjacent space. At forty-seven years, he's doing well to avoid losing the onset of a middle-age skin-versus-gravity war.* MacGillivray took in Pendle's brown eyes, healthily clear, unblemished with bags. *He sleeps the slumber of the untroubled; they live amongst us.*

'I am DCI MacGillivray, and this is DI Crozier. How was your journey, sir?'

Pendle ignored both their proffered hands and the conversational icebreaker, instead asking, 'Do we have a meeting room?'

'This way,' said MacGillivray, trying to ignore Crozier's raised eyebrow but also irked by the man's attitude.

The five rode the lift in uncompanionable silence

to the same floor as Grey. The 3rd-floor digit on the lift panel lit up, confirming their arrival.

MacGillivray led Pendle to a commandeered visitor's waiting room. A disgruntled cohort of visitors loitered, aggrieved at their eviction. Now curious and increasingly nervous at seeing the uniforms. *Animals sensing predators.*

'Thank you, DI,' Pendle said to Crozier. 'Your attendance is not required.'

MacGillivray threw her a sympathetic glance.

Crozier exited, teeth clenched, delivering a borderline brusque shut of the door behind her.

Pendle then told his two operatives to stand guard outside Grey's room before turning to face MacGillivray.

'We have noticed your *heroic actions*.'

MacGillivray didn't think the irritation in Pendle's eyes matched the compliment of his words.

'We live in precarious times, and confidence in the Government and the police services is fragile, so anything we can do to build trust is paramount. You can help by closing cases as quickly as possible.

And you can ignore the *Grey and Holroyd* business; consider them within NSSA jurisdiction. In strictest confidence, I will say that they are working undercover, so any interference from you will jeopardise the entire operation.'

*NSSA staff? Is the NSSA the third party? How many incidents were they responsible for? And why? He could scarcely believe that his hunch was taking on flesh.*

Pendle continued, secure in not needing an acknowledgement from MacGillivray.

The DCI interrupted anyway, 'What operation is that, sir?'

'That information's on a need-to-know basis. And

you can guess the answer, DCI, unless you join us. And that will depend on how the next few days go.'

'You want cases closed quickly, but they may not be so cut and dried,' persisted MacGillivray.

Pendle replied, 'There isn't enough time or resources to chase them down. Besides, most of them are usually guilty of something.' *People's rights swept away with Pendle's airy dismissal, acceptable collateral damage in the battle to balance a budget to gain power.*

Pendle stood, a sign of the meeting's end, his instructions relayed.

'But...'

Pendle, patience dwindling, scythed him down, 'It won't look good for your budget prospects or your role opportunity with the NSSA if you continue to fail.'

'Yes sir,' said MacGillivray, not seeing any value in antagonising him further.

Pendle said. 'I hope you fare better than your predecessor, Alderson.'

MacGillivray felt the weight of implied criticism. He also felt the implication of Pendle's veiled threat. *Simon Alderson had died of a heart attack–am I paranoid to be reading too much into Pendle's words? Had Alderson uncovered something that he shouldn't have done? Had the NSSA killed him?*

'Do your part, and we'll all get out in good shape. Besides, MacGillivray, haven't you lost enough?'

A flare of anger lit up within him. Through gritted teeth, MacGillivray said, 'Understood, sir.' *How dare he invoke the deaths of his family to motivate him?* He felt an almost blinding need to grab Pendle's throat and throttle him. MacGillivray gripped his chair, helping him to fall on the side of non-violence.

Pendle laid a hand on the door handle, turned and

said, 'Let me clarify; you and your team can leave. We'll interview Grey when he revives. You have enough to do.'

---

'I SEE a couple of Nazi finishing school candidates have replaced PC Hanbury at Grey's door.' said a returning Crozier, bearing hot drinks.

'Let's keep this professional,' said MacGillivray. *And how far does the NSSA's surveillance go? Was this room bugged?*

'How did it go with Pendle?'

Suspicious of the room, he signalled for them to go into the corridor, much to the bemusement of Crozier.

'Pendle's warned us off checking on Grey and Holroyd, as they are *working undercover*.'

'Oh, right?' She held his gaze, reading between MacGillivray's lines. Absorbing the implications. 'You said they'd been shooting at each other.'

'Alliances have changed, it seems.'

'Did you tell him what you saw?'

'No.'

'You don't trust him!'

'The jury's still out. But we're under pressure to close outstanding cases. If we don't, it will be the end of my career and the station's closure.'

'You don't look like you're prepared to toe the line.' MacGillivray held a neutral face.

'Well, whatever you decide, count me in.'

'You don't have to commit career suicide, too, you know.'

'I've got to live with myself when this is all over. And here's another question, before you waste time trying to talk me out of it, where has our friend Miss Holroyd gone?' asked Crozier.

'That is a good question.'

Another unpalatable question occurred to him: *had Crozier leaked information? Why would he think that? Because of the exposure of Grey, that's why.*

*Could he trust anyone with so much at stake?*

# 25

# Holroyd

Holroyd had made it to the hospital car park in time to see MacGillivray and DI Crozier leave the building.

*Why were they leaving Grey? Wouldn't they wait for him to wake up if he's not conscious?* Whatever the case, she figured she had scant time to deal with Grey.

*And what about the detective MacGillivray; had he been coming to question her at the apartment? He must be getting close to the truth. Grey had said he was ineffective, a defective detective; he'd quipped in a rare moment of humour, but MacGillivray was proving to be anything but —another misjudgement on Grey's part.*

*Focus,* she told herself. *First, she had to get to Grey.*

Holroyd found a fire door at the back of the building. It had been wedged open while hospital orderlies took a break. They were enjoying the easy companionability of their smoker's club, even within the cool draft of a deluge of rain. They were talking about the shooting and fortunately mentioned that Grey is on the third floor.

She flashed her ID at them and warned them if they're caught smoking on the premises again, they'd be fired. She accessed the staircase, leaving a chastened and, importantly, unquestioning hospital staff behind her.

HOLROYD ENTERED Grey's room with the invisibility of a white coat, mop and bucket she'd easily sourced from a convenient storeroom.

She turned the corner in the corridor. Two bulky characters with gun bulges are standing outside, she assumed, his door: *Pendle's NSSA cleaning squad were here already. So why not declare herself to them? Because she needed to get this kill for herself, to prove her worth, and to stay in the NSSA.*

Holroyd tried to shut the door, but one of the two guards held it implacably open. *Killing Grey would not happen yet.*

*But she had an idea.*

GREY LAY ASLEEP, chained to the bed, head and arm bandaged. She felt a contradictory concern for him.

Holroyd ran the mop across the floor, superficially scenting the room with disinfectant, lending authenticity to her role.

The view of the window offered an outline of the Town Hall clock tower through a half-hearted rainfall, along with the black metal rails of a fire escape.

Holroyd discreetly lifted the handle on a window to leave a gap, then left Grey's room unchallenged. *If Pendle found out, he would have them hung, drawn and quartered for letting her in and out so easily.*

HOLROYD'S FOOTSTEPS left brief indents in the

rivulets of water running down the fire escape, and from this high up, the wind felt more prominent.

She sneaked a look into Grey's room and saw him asleep. And importantly, the door to his room is shut. *Now, she could kill Grey, uninterrupted.*

Holroyd climbed in, deftly dropping onto the floor.

But she still woke Grey. He looked blunted by sleep and painkillers, but soon sharpened with cortisol at the sight of her gun. He sat up as if that act would ward off his doom, yet Grey didn't even call for the guards, probably because once the guards knew he was awake, they would have access to him and be able to deliver whatever unpleasant fate they had in store for him.

Grey rattled the handcuffs he's tethered to the bed with, looking trapped and resigned.

Holroyd looked for a pillow to use as a silencer, determined to get this job over with as quickly as possible and to avoid the complications of what she felt for Grey.

She stopped, alerted by an argument escalating outside Grey's door.

'I need to see my operative.' She recognised that voice as Pendle's.

'I am Mr Hughes, and I insist you allow him to rest.'

She hoped this Mr Hughes would keep Pendle preoccupied long enough for her to put a bullet into Grey's head.

'We don't have time for this. It's a matter of national security that I speak to him,' said Pendle pompously.

'He is my patient, and I have a duty of care to ensure he gets time to recover.'

'Then speak to the Minister for Crime, Policing and the NSSA. I am sure Ella Peterson will have something to say on the matter,' said Pendle piously.

'This is outrageous!' said Mr Hughes, though she registered doubt in his voice at the floor-denting name drop of such a high-ranking person.

On hearing the door handle turning, Holroyd hid behind the partition curtain as Pendle barged in, his immaculately uniformed presence filling the room.

*Why didn't she declare herself? Because she didn't trust Pendle, that's why.*

Pendle stood at the end of Grey's bed, appraising him, and without preamble said, 'When are you coming back?'

'Almost there, sir,' said Grey, *with remarkable control, considering.*

'That's good news, we could use you.'

'Is that why you ordered Holroyd to *take care* of me, Sir?' asked Grey.

Pendle threw his arms open in a gesture of generosity and forbearance, reaching for an avuncular smile hoping for soft acceptance from Grey, instead hitting something hard: like a spade hitting rock.

'A misunderstanding, Grey. You'd made a mess. But I am now aware of mitigating circumstances, and we are here to help you clean it up.'

'*Mitigating circumstances, sir?*'

'The boy, the detective, the weather, all bad luck. Holroyd has been told to step down; you needn't be concerned by her.'

*You lying bastard,* thought Holroyd, now feeling justified at electing to hide.

'Update me on the mission.'

Holroyd took in Grey's strung-out story. *Grey is playing for time by trying to get Pendle to talk. He thinks he's going to die either by his or my hand - but he wants me to hear what Pendle might give away.*

Pendle paused heavily, then spoke. 'We are dying as a nation under an avalanche of public debt. We must make hard choices to cull the population of those draining our resources to unburden the economy. And, with a rising wave of organised crime threatening to overrun the country, who better to cull than the organised crime syndicates and other criminal elements when they present themselves? Did you know that one in nine of the population is now involved in criminal activities? We can spare the overloaded judicial system from dealing with this national threat.'

'Who would be targeted next?'

'Let me provide a spoiler alert, Grey: later, we'll work our way through other wasteful elements of society, which will be for the greater good.'

Holroyd had felt some pride at aiding the quelling of a mass uprising of crime and dissonance because they threatened to destabilise the nation. *But who else does the NSSA define as a burden to society? It looks more like a cold-blooded culling of the weak and vulnerable* of *the desperate poor looking for ways to survive – by criminal activity or otherwise - all for political expediency for the incumbent government and their wealthy donors to hang onto power.*

She and Grey had been willing instruments to carry out *Project Deadhead but now* Holroyd could comprehend Grey's internal turmoil–*he had got closer, quicker, to the reality than she had. He had been struggling, feeling himself eaten away by what they were doing, which made him a better person than her.*

She could just see Pendle; he looked smug, confident that he had vindicated everything he and the Government had done and what they would do. *If Pendle felt comfortable telling Grey this, to boast about Project Deadhead, then Grey is a dead man.*

'It's about keeping the power with the few,' said Pendle, to keep the economy in the grip of the few, to bend the masses to the will of the few.'

Holroyd, holding her breath, squeezed her gun, knuckles white.

*This Government thinks it is morally superior. Yet when the pressure is on, they are equally bad — matching any other toxic, self-serving entity.*

*Shoot the bastard,* she urged herself.

Again, taking counsel from her inner self, Holroyd waited, suppressing the impulse to kill.

'Who is responsible for Project Deadhead?' asked Grey.

*Good question, Grey.*

'Peterson gives me the clay of instructions. I am the artist that creates the sculpture,' said Pendle. *The arse,* she thought. *He means Ella Peterson: The Minister for Crime, Policing and the NSSA - tipped for the PM hot seat. She sanctioned Project Deadhead.*

Mr Hughes is indignantly demanding to be let in, barred by the two NSSA operatives *yet remarkably undeterred by their guns.*

'Let him in,' said Pendle.

Face redolent of freshly cut beetroot, Mr Hughes said. 'It appears you have friends in prominent places. However, I insist that if you don't leave my patient to rest, I'll hold you responsible for any degradation in his health.'

'Don't worry,' Pendle delivered his words leavened with contempt, 'I am finished here for the time being.' Then he turned back to Grey; suddenly all faux visitor cheer and said, 'Rest and let's see you back in service soon, Grey.'

Hughes, Pendle and his sidekicks exited, but Holroyd, who rushed to put an ear to the door, could still hear Pendle addressing the two NSSA operatives.

'When it is appropriate, finish the job. If Holroyd appears, finish her, too.'

She turned, locking eyes with Grey, his jaws muscles clenching, agitated with questions.

She could answer them.

*No, she would not kill him now.*

*Yes, she's going to help him escape.*

*Yes, we're in deep shit.*

All conveyed within a blink.

He nodded his pain-filled, relieved understanding. *He'd not die–at least not right now, and the band is back together again.*

Utilising a small multi-tool, Holroyd freed Grey from his metal bonds. She helped him out of bed, moving it in front of the door. She put the wheel locks on as one of Pendle's guards turned the door handle, met resistance, and banged against it.

Grey and Holroyd made their way out onto the fire escape. Grey dressed in only a hospital gown, rain hitting him without due care and attention.

The NSSA operative delivered a solid barge to the door, shaking the bed back a few inches. *They had little time to get away.*

Holroyd helped Grey drug-stumble down the fire escape, hands dislodging globules of water gathered on the metal handrail. As they reached the bottom, bullets rained down from the two guards, hitting the tarmac just short of them.

Holroyd turned and fired a reply, forcing them away from the window.

She led Grey through the rain to her Polo.

'There are clothes on the back seat.'

'Thanks,' said Grey, his pallor matching his name.

Grey grabbed at the neatly stacked black jeans, a white fisherman's jumper, and a black woolly hat and jumped in the front passenger seat.

Holroyd stabbed her finger on the ignition button. 'Come on, start.' she bellowed, unhelpfully smashing her fist on the plastic dashboard.

'Are you sure you're pressing it for long enough?'

'There's always time for *mansplaining*, right?' Holroyd couldn't tell if Grey winced at her comment or his pain.

She pressed the button again, holding it; this time the engine graciously deigned to start.

Holroyd took a left out of the car park, the shock absorbers working hard as they hit hidden cavernous potholes under generous pools of rainwater.

A QUARTER OF A MILE ON, she swung the silver motor to the side of the road.

'What are you doing?' asked Grey.

'We need to get on the offensive, which means getting to Pendle. We'll not have a better chance than now.'

'Bold. I like it,' slurred Grey.

'But first, let's deal with our pursuers.'

Holroyd got out, crouched, waiting.

They came soon enough, racing towards them.

Holroyd stepped into the middle of the road.

She pointed her gun.

The black Land Rover's driver took this as a cue to speed up. She could see the passenger leaning out of the vehicle window with a gun aimed at her, the gunfire audible above the engine's roar.

Holroyd's first shot missed the right front tyre, while the second didn't. The Land Rover folded to its right, the driver fighting the steering wheel for control and failed as the vehicle hit the kerb and flipped. It had enough speed to complete a half spin before landing on its roof; a brick wall acted as an efficient brake, bringing the NSSA transport to a rude, metal-scraping halt.

Holroyd ran over to investigate the wreckage.

The driver's skull, an eggshell crushed, providing a gush of blood and grey brain matter that augmented his dark blue attire.

A groan came from the driver's colleague. *He was still alive.*

Hawham High Road stood still in strained silence, the air filled with the ticking of a cooling engine. Then, the tension resolved itself in an ignition of petrol and the whoomph and climax of fire, Holroyd catching the cloying smell of burning fuel permeating through the drizzle.

The passenger stirred, primal fear of fire gifting him energy to struggle. *He's trapped, and he is beginning to burn.* He shouted at her to finish him as the fire started about its task of consuming him, his shouting rapidly migrating to a scream.

Holroyd considered leaving him to his unpleasant fate but relented and shot him in the head. The man's head fell into a pain-released repose as the smell of seared flesh assailed her nostrils.

'If governments had kept their promise to phase

out fossil fuels, he might still be alive,' said Grey when she returned.

'Funny. Let's get Pendle.'

THE ROAD REDUCED to a single straight carriageway. Along its side, the looping telegraph cable rose and fell between poles, giving the optical illusion of being alive. The headlights of their car shone a round beam ahead, making the trees dance in and out of vision.

Ahead is Pendle's vehicle on the deserted road.

'Hide,' said Holroyd to Grey, who obediently dropped out of sight. *Pendle shouldn't see him, or he might smell a rat.*

Holroyd overtook Pendle's Land Rover. Then she sounded her horn, attracting Pendle's attention, pointing to the side in a request for him to pull over. Looking simultaneously surprised at seeing her while barely masking his irritation, Pendle followed Holroyd into the next layby.

Holroyd got out and walked to Pendle, who had lowered his window.

'What do you want, Holroyd?' he said testily.

'Apologies for interrupting your journey. We're hoping you could help us, sir?'

'We?'

Then he saw Grey appear.

Pendle's eyebrows lifted, betraying a catalogue of questions.

He chose to start with, 'What is Grey doing out of the hospital?' and quickly followed up with his second choice of 'And why haven't you bloody killed him?'

'I heard everything you said at the hospital.'

Comprehension and trepidation, now Pendle's two new bedfellows, are filling his face with a red flush of rage and fear.

'Now, get out,' said Holroyd.

Her gun, she hoped, would dissuade him from thinking he had a choice.

'Now listen, if you make the right decision, Holroyd, you can still come out of this intact.'

A lorry passed their little theatre of drama, throwing up water in their direction. 'It's you that might not come out *intact*. Now get out and turn around.'

'Fuck you, Holroyd. You insubordinate...'

Holroyd slammed her gun into Pendle's face. Blood, given free rein, escaped from the nose, dousing the front of his coat.

'You...you... shit. You...!' She hammered her gun but into his face again, his cheek opening, offering forth rich blood. He fumbled delicately around his face with his fingers, surveying the damage, grimacing at his findings.

'Do as I say, or it'll be a bullet next.'

Pendle obeyed, one hand on his face, the other pulling himself out. She bound his wrists with a plastic tie.

'Get in the back.'

Pendle sullenly complied.

Then she punched him on the jaw, knocking him clean out. His yielding face felt good on her knuckles.

Holroyd spotted a phone box at the end of a row of houses, a last outpost of Hawham suburbia.

'We need allies, so I will call MacGillivray, as he might be our only hope.'

'Won't he arrest us and hand us over to the NSSA?' asked Grey.

'I think he has some idea what is happening, and if we can confirm his findings, he might help us. And he might need our help when the NSSA visit.'

'It's a long shot, but okay.'

She entered the phone box and called MacGillivray.

# 26

# DCI MacGillivray

MacGillivray and Crozier returned to a palpable buzz around Hawham police station at the revelation of Grey's identity. MacGillivray thought he could even interpret the looks from Hanbury and Mason as respect. Not that he had time to revel in any adoration as his phone rang as he entered his office. MacGillivray picked up the receiver and then mouthed to Crozier, *its Holroyd.*

*How does she know his direct number? Silly question, it's hardly a major obstacle for the NSSA to find that out.*

The call didn't last long.

When MacGillivray put the receiver down, he said to Crozier, 'Holroyd and Grey want to hand over Pendle at the market square. They will tell us as much as they can in exchange for help.'

'Now Holroyd and Grey are *not* shooting at each other. It's hard to keep up with those two,' said Crozier. ' I can't wait to hear what they have to say.'

'Let's find out,' said MacGillivray. Within minutes, he, Crozier, Hanbury and Mason are kitted out in bulletproof vests and weaponry and heading out to meet them.

# 27

# Holroyd

Holroyd drove them towards the market square. Grey, in the passenger seat, is riding shotgun. *He is bleeding and looking even paler. Grey could do with attention,* she thought. Groaning sounds came from the reviving figure of Pendle. *So could Pendle, but to hell with him, he can suffer.*

'We have company,' said Holroyd, seeing a large Ford in the mirror; the sight triggered another dump of fight-or-flight chemicals, bathing her senses.

Ahead, parked on the curve of the suburban road, is a black Land Rover Defender. The Ford behind them, she noted, is slowing to a standstill, keeping its measured distance. And, with Sussex stone walls on both sides of the road, they are trapped.

'Fuck, we're trapped,' said Holroyd as she slowed to a halt.

A man appeared from the Rover ahead and walked towards them, stopping about thirty metres away. He shouted, 'You've got one minute to give Pendle up. Do it, and I give you my word you can go.'

'That's Sam Marsh,' Grey said to Pendle. 'I killed his boy, Tommy. Because you misled us.'

'You took very little persuading to kill *children,*' sneered Pendle.

Grey's temper flared, punching Pendle with his left hand, getting a decent hit, even from behind his seat. Blood trickled from where Grey's ring had caught Pendle's other cheek.

Pendle's petulance blended with pain and fear, the latter two now babblingly dominant, as the big man was reduced to pleading, 'If you give me to these thugs, where's your bargaining power? With them or the police? You need me to get you out of this!'

'If you haven't noticed, we're outnumbered and outgunned, and we could get roasted, too,' said Holroyd, nodding towards Marsh.

Sam Marsh held a bottle with a cloth hanging from its neck. He lit the Molotov cocktail and threw it.

The bottle rose and fell, landing short of Pendle's Rover.

Only Pendle gasped.

It burned, urgent flames augmented by the dark, its fire consuming the tarmac through a sheen of wet and flickering irregularly on the flint walls around them.

Marsh had a second bottle; a baby couldn't have been handled better.

'Thirty seconds left,' shouted Marsh above the bullying crackle of flames. 'Then this one hits your Rover.'

'Sorry, Pendle,' Holroyd said. 'We don't have a choice. Out.'

Despite the grim nature of their plight, Holroyd considered Pendle's look of misery and his brutal fall from his cosseted world as comical.

'How can you be enjoying this?' he said, alerting her that she had a smile.

Aided by Holroyd's insistent prod from her gun, Pendle vacated the vehicle.

'You're going to regret this,' said Pendle, barely convincing himself of the power of his threat, much less Holroyd.

Grey followed, and Holroyd recognised the tension in his body language, reading his intentions. *He's ready to welcome the fight.* Because Holroyd knew Marsh would break his promise when he got what he wanted because, logically, he couldn't let them live.

---

'YOU'RE MARSH?' said Pendle. Even though the rain had stopped, it left a crisp cleanliness, and his teeth chattered with the chill, fear, or perhaps both. 'When we're done here, you'll have it all, free reign over the town.'

'It's good to meet you, Pendle,' said Marsh, his sarcasm a poor mask for his grief and hunger for revenge. 'But, in the interests of providing full transparency, we ain't good news for you as I am the father of a boy you killed.'

'Tommy Marsh?' he said, nodding towards Grey. 'He carried it out.' Holroyd could almost hear Grey's body being mangled under a bus; the one Pendle had thrown him under, Pendle's reeking desperation disgusting Holroyd.

'Good to see loyalty isn't dead,' said Marsh, eyes now locked on Grey, receiving nothing but silence.

'You killed my boy?' he shouted at Grey. 'You're not fucking denying it?'

'Yes, I killed your boy,' said Grey.

'But I am guessing Pendle issued the order,' said Marsh, to a nod from Grey. 'At least you're owning up to it. Unlike this fucking coward,' said Marsh, looking at Pendle. Then Marsh took two steps forward and spoke. 'Why *did* you have Tommy killed? Didn't I do what the NSSA wanted?'

*What The Fuck? What the hell else is going on here?* Thought Holroyd.

Pendle looked resigned to his fate, but then a little of his old arrogance surfaced. 'It was supposed to be a simple job, and you messed it up by killing MacGillivray's wife and child. Why would we reward such incompetence?'

*Had he organised a raid on a senior detective's home by using organised crime thugs?* Thought Holroyd. *But is she any better? When boiled down to it, government agent or not, she's a gun for hire, killing on the same instructions from the same person. She's in no position to command the moral high ground.*

'You get the big bucks to give the orders, so you're going to take the consequences, fucker,' said Marsh, nodding to his son Troy, a signal to put Pendle into their Rover.

Troy laid a hand on Pendle's disfigured face, shoving him towards their car. 'Move,' he ordered.

'Ow. Mind my face, you cretin.'

'You won't have to worry for much longer,' said Troy, his counterfeit smile tugging skin lines taunt under the hard orbs of his eyes.

'Wait, what DO you want? I can make it happen,' said Pendle, his voice quavering, the sighting of his backbone only a mirage.

Marsh nodded to Troy to stand down, then said, 'You know, you people have had the power of life and death over others. Then, when confronted with the consequences, out here in the wild, you fold. You all make me sick. You blackmailed me into raiding Mac-

Gillivray's place and promised the town to me once you closed the police station, a deal you're doing a decent job of pulling out of. So why should I believe you now?'

'He's a piece of shit. Don't waste your time.' Troy pointed his gun at Pendle.

'Put the gun down, Troy,' said his father.

Troy held the gun in situ. Then smiled. One, Holroyd is sure, comes before pulling a trigger.

'Troy, I am warning you,' said his father.

Troy reluctantly dropped his gun arm, winking at Pendle.

'You see my problem? You can barely control your son, much less this town,' said Pendle. 'And can you really keep your mouths shut? Prove why I should trust you, and I'll pay you two million pounds,' said Pendle.

*Bold and assertive, she almost found something to admire in Pendle. But would his gambit work?*

Marsh had stopped, contemplating Pendle's words.

'Seriously? You reckon you can trust this double-crosser?' asked Troy.

Marsh took a gun from his pocket and pointed it at his shocked son. 'Will you, for once, shut up and let me think?'

# 28

## DCI MacGillivray

A fire blazed, casting a woodcut image of three vehicles and a small cluster of people against a wall.

They'd all stopped in mid-motion, looking MacGillivray's way.

It felt like something from medieval times, as if he'd caught witches at work in their coven stead.

'There's Grey and the chimeric Holroyd,' said MacGillivray. 'Though it looks like they are in the process of losing Pendle, their bargaining chip.

MacGillivray thought Shakespeare had it right when he had written that the eye was *the window to the soul*, because when he gained eye contact with Pendle, he saw naked terror turning to undiluted relief.

'That's Troy Marsh bundling him into their car,' said MacGillivray. 'We appear to be in just in time.'

Not in time to stop Troy Marsh from punching Pendle, knocking him clean out.

---

MACGILLIVRAY HAD MADE ANOTHER MISTAKE.

He should've considered how James Leigh might

react to seeing the suspected murderer of his family enter the police station. Because he leapt at him, shouting, 'You're a murdering bastard.'

MacGillivray just managed to get between James and the hand-cuffed Marsh, though the lad still landed a wild punch on Troy, who had come to his father's aid.

Troy clubbed James in the face with his hand-cuffed fists. The boy collapsed to the ground, stunned.

'Did you have to do that, you bloody savage?' said MacGillivray as he held James in case he tried to hit out again.

'It's instinct... self-defence,' said Troy Marsh, looking pleased.

MacGillivray scowled. 'I should charge you with actual bodily harm.'

'But you'd have to do the same to the boy.'

*He was right. And he would not do that.*

'Now listen, kid,' said Sam Marsh, addressing James, holding his clenched hands up to ward off any other attacks from the boy. 'I'm sorry to hear about your folks. I know there's been misunderstandings and friction between our families, but I swear we had nothing to do with their deaths.'

'Yeah?' James snarled, blood pulsing from his nose, mixing with pain-induced tear trails. 'Prove it.'

'We've been set up, son, believe me. Someone is causing trouble between our families. I spoke to your old man yesterday, and, believe me, he thought the same.'

Despite his fury, James stared at Marsh as if divining for some truth in what Marsh was saying.

'It's possible, James,' said MacGillivray. 'And I am going to get to the answers.' MacGillivray stared hard at Pendle and Holroyd, who both looked immutable.

While Grey, at least, had the decency to look uncomfortable.

James rose, hands clenched, struggling with this new information. MacGillivray kept a restraining hand on him.

James faced MacGillivray. 'Why should I believe what you say? You're the bloody police, and my dad says we buy you.'

'Not all of us, James,' said Crozier.

Holroyd spoke. 'You should know I...'

MacGillivray interrupted Holroyd. 'He doesn't need to hear it right now.'

'Know *what?*' asked James, eyeing Holroyd curiously.

'Let's get it out in the open,' said Holroyd. 'I killed your family, James.'

'For heaven's sake, hasn't he gone through enough?' asked MacGillivray.

James had already sparked, trying to attack Holroyd, but this time, MacGillivray held him by the arms, James' jaw working in grief and anger.

'Why did you have to kill them?' he demanded.

'Because the government is trying to stop the escalating crime wave,' said Holroyd.

*It's cruel to hit James with another brutal revelation. But maybe Holroyd is right; perhaps she's the one person here giving the boy what he needed: the truth.*

*And it also meant that he didn't have to consider James a possible suspect.*

'At least someone is telling the truth,' said James bitterly.

Grey sighed, adding. 'And I killed Tommy Marsh and Eddie Bridges to spark a war between the Marsh and the Leighs.'

*That confirms my suspicions,* thought MacGillivray.

'The government orders the execution of people, of *teenagers*. And you *do* it?' asked James, his words directed at Grey and Holroyd.

'He's got a point; that's a good question,' said Crozier.

'We've carried out missions to kill adults who were a threat to the nation. Is there a difference if the threat comes from someone younger? And, honestly, is it so wrong to clear societies' scum out?' asked Holroyd.

'What did you fucking say?' Marsh moved, ominous but impotent. His strained handcuffed hands held up, 'Tommy's not fucking *scum*.'

Grey and Holroyd snapped into action, raising their guns.

Marsh's face had turned puce, a heart attack perhaps a step away, repeating, 'Don't call my son scum,' stepping back nonetheless.

James broke the moment, looking at Grey, 'You... you're the man who tried to kill me at the river. You shot the guard!'

'Listen...' said Grey.

James interrupted him, saying, 'Yeah, it's always someone else's fault, right?

And you...whatever name you call yourself...aren't you the government? Aren't you supposed to protect us?' Then, looking at MacGillivray, James said, 'And the police - you take bribes - who is *the scum* here?'

'Yeah,' said Troy Marsh, nodding, 'Fucking right.'

James slumped into a chair, dropping his head into his folded arms.

Fraught, dense silence held the room.

Then Holroyd broke it, saying, 'So what happens now?'

'I don't know,' said MacGillivray, rubbing his eyes.

'Pendle and The NSSA are preparing to silence anyone who knows about it,' said Holroyd.

*She is not one for dressing up bad news,* thought MacGillivray.

'You mean they're gonna... kill us all?' asked James, 'Shit, can't we tell someone?'

'We're trying to do that,' said MacGillivray.

'But you failed so far, is that what you're saying?' said James.

'The Victorians used to say children should be seen and not heard,' said Holroyd.

'And shouldn't government agents not be killing its own people?' said James.

'Touché,' said Crozier

'Enough,' said MacGillivray. 'Let's at least lock these suspects up.'

MACGILLIVRAY AND CROZIER escorted the Marshs to the cells.

Troy asked, 'Bloody hell, can't you raise the heating a bit? It's chilly down here.'

'You might've noticed the country is on an economy drive. And the police are trying to fund the fight against an increasing crime wave. Isn't it poetic justice that criminals should feel the cold here, sir?' said Crozier.

'We're innocent until proven guilty.'

'Quite so, sir,' said Crozier, giving Troy a withering look while she increased the heating by the wall-mounted controller. 'Are you comfortable now? Perhaps I could get a few scatter cushions? They'll come in handy for your coming years in confinement. You know, for the kidnapping of a senior government officer.'

'Yeah. A murdering, blackmailing, senior government officer,' said Marsh senior.

For all Marsh's bullish words, MacGillivray saw a prophecy of hard prison years foretold in his eyes.

'Look, if I give you some information, will you go easy?' asked Sam Marsh

'Dad! What are you doing? You're not grassing, are you?'

'Your observance of the criminal code of behaviour is endearing,' said Crozier.

'Fuck off,' said Troy. 'I hear your station is gonna get shut down. You lot will be out on your arses, and then hunting you down will be fun.'

'Button it, Troy.' said his father.

'It'll be fun, for sure,' said Crozier. 'A town without the law. It's a blade that'll cut both ways, right?' Troy flinched at her comeback, leaving a gap in the conversation which MacGillivray quickly filled. 'I can't promise anything, but it might help your case. What have you got?'

'Even if I provide evidence, what will you lot do? It's Pendle, right? He's one of your own. You'll close ranks, and he'll never go to court.'

'Said without a trace of irony.' Even with MacGillivray's best efforts, Marsh had escaped justice on numerous occasions.

'Yeah, and if I'd not been useful to the NSSA, I'd have been sent down. I'm not proud of some of the things I've done, but it's all for the survival of my family. But *the system* has used us, and now, it is killing us. It Killed my boy.' His voice hitched. 'And I want justice.'

'So do I, Marsh,' said MacGillivray. 'So do I.'

*MacGillivray had believed that compassion and empathy were the keystones that kept us civilised. And*

*Marsh's particular brand of compassion was why he protected his family.* But *they'd both failed to do that - they had that much in common.*

*And Marsh was right about the country, about the system. Hooligans were not just running the streets. Some, like Pendle, were running the Nation, enabling the elite to get a stranglehold on everyone and everything. And maybe they had always done so.*

*So MacGillivray couldn't condone Marsh's behaviour, but he could understand it.*

Crozier broke his spell. 'This socio-political musing is all very fascinating, but you're not giving us anything to work with.'

Marsh scowled. 'She needs to know her place; she's a cheeky cow.'

'Do you think he means I should return to where I came from? Isn't that a shade racist?' said Crozier.

'No, I believe he meant that a woman's place is in the home. Perhaps it's Marsh that's in the wrong place.'

'Yes, he should be back in the middle of the last century,' said Crozier.

Marsh, red-faced and perplexed, said. 'Alright, keep your hair on. Pendle started singing once he thought his life didn't have long to run, if you catch my drift. Not that I am admitting to anything. He said he'd instructed the two NSSA agents to do the dirty work. He'd used a list of names he'd been supplied from NSSA field intel.'

*That information supported Holroyd and Grey's version–and someone had supplied a tally of names to Pendle from Hawham police station and given a case for their execution. The list that Ash had taken from Grey and Holroyd's apartment.*

*So, who had generated that list?*

*And could Project Deadhead be happening outside of Hawham, too? How big is this?*

'What about us?' asked Marsh. MacGillivray didn't answer as, deep in thought, he left the cell with Crozier in tow.

'What about a deal?' shouted Sam Marsh.

'I already knew all that. I want more,' said MacGillivray over his shoulder. He paused, then asked. 'How did you know Pendle would be dropped here?'

Thrown by the question, Sam Marsh looked sideways as if for a plausible lie.

'I'd heard a whisper about Pendle being picked up for his part in Tommy's death. You can't blame me for wanting to get him, right?'

'Maybe not. And you can't blame me for wanting to know who's leaking information. Speak now or forget about a deal.'

Marsh looked to be struggling for a creditable lie, caved in, and said, 'Jacobs.'

Troy looked disgusted. Perhaps *the last remnants of awe for Troy's childhood god had gone?*

MACGILLIVRAY GATHERED CROZIER, Holroyd, and Grey in his office.

He addressed the NSSA agents, 'I ought to be arresting you both for murder.

But I have a hunch we'll need your professional skills soon enough. And, with you holding guns, you are unlikely to come in quietly, anyway. But let me make this clear: I promise nothing when this is over. So, are you in?'

'No bullshit, I like that. We're in,' said Holroyd, looking at Grey for confirmation.

Grey nodded in agreement. He looked in discomfort, the bruise to the side of his head discolouring, his bandage was showing signs of leakage, and he had sweaty signs of infection. *Grey needed treatment and rest but would not get it for a while, that was certain.*

Catching MacGillivray's look, Grey said, 'I'm fine. And I owe you some thanks. You could've left me in the car when Holroyd came for me. I wouldn't have blamed you. You're a good man.'

'I hope I don't regret the decision,' said MacGillivray.

Grey nodded, 'Thanks, anyway.'

'So, what can you tell me?'

Holroyd spoke, 'We work - worked - for the NSSA. We were briefed about a new initiative called Project Deadhead and then given a file of targets. It comprised crime ring leaders and allied members of the syndicates who need to be...'

'...Dead headed?' asked Crozier.

'Yes,' said Grey. 'Pendle told us it's for the greater good of society. Only it transpires that it is solely for political expediency so that this party can stay in power and keep the backers in positions of influence.'

'But you two have opted out?' asked Crozier to nods from Grey and Holroyd.

'How big is Project Deadhead?' asked MacGillivray.

'We were assigned to teams of two, allocated a town with prescribed undercover jobs with designated criminal elements as targets. We had to ensure the deaths appeared as accidental as possible. Then, once we had completed our kill list, we'd move on to a new town. We don't know precise numbers, but there are teams working across the country.' Holroyd gauged their reaction. 'You don't seem surprised,' she said.

'We've found clues,' said MacGillivray.

Holroyd nodded. 'We've not been as clandestine as we'd like.'

MacGillivray caught the sharp look directed at Grey.

'Why were you trying to kill Grey?' asked MacGillivray.

'Grey had messed up too many times. Pendle ordered me to *'retire'* him from the project.'

Grey bristled but said nothing.

'Then I heard Pendle's give a little speech to Grey at the hospital and knew we'd been played.'

'Why *didn't* you declare yourself to Pendle? Why hide?' asked Grey.

'Intuition, luckily for you,' said Holroyd to Grey, then facing the others, said, 'We're not naïve; we know we've done borderline things for the government by justifying those actions for national security. To be fair to Grey, he had been questioning the legitimacy of our actions for a while.'

MacGillivray glanced at Grey, noting the surprise on Grey's face. *Perhaps at being understood better than he understood himself?*

Holroyd took a pause and said, 'We're screwed. But we can try to redress the balance before the NSSA gets to us.'

MacGillivray noticed Crozier's face softening, perhaps realising that even assassins are human.

*These revelations, making our choices grow limited, have brought us close to the brink of disaster. And here we are wasting precious time deciding whether to trust each other,* thought MacGillivray.

'If you were going to shoot us, you would've done it by now. And everything you have told us seems to fit the evidence we've gathered. And without your help, we don't have a chance of getting out of this alive, anyway.'

'Trust me, the odds aren't great, even with our help,' said Grey.

'Thanks for the pep talk,' said Crozier.

Grey threw her a weak smile.

'We need to speak to Pendle. But I want to let him stew first while I make a few phone calls,' said MacGillivray.

---

PENDLE HAD REGAINED consciousness in the cell. He had tissues sticking out from each nostril of his swollen face, making him look faintly comical. He took the coffee offered to him by MacGillivray. 'I want Marsh sent down for life. And as for that bastard that hit me, if I get my way, he'll be lucky to see out the day.'

'I think it might be prudent to follow some vague process of law, sir.'

Pendle looked at MacGillivray and said, 'You may have a point'.

Perhaps gratitude was a new sensation for him, but he managed to say, 'And thank you for saving my life, MacGillivray.'

Then he stood up, shrugged off his pain as much as he could and attempted to resemble authority by saying, 'But I would like to vacate my temporary accommodation and have a full briefing in your office now, DCI.'

'All in good time, sir. First, we need to ask you some questions.'

'Ask ME some questions? Are you insane, man?' said Pendle, his gratitude dissipating fast.

Pendle tried to push past MacGillivray. But MacGillivray held firm, standing in his way, while Crozier stepped outside the cell, locking the door and turning up the red dial of Pendle's face.

'What the *hell* is going on? You know the role I offered you may depend on how you perform now?'

MacGillivray could feel Crozier's questioning stare of why he hadn't mentioned this before.

'Yes, sir, I do, but it would be a dereliction of duty not to detain you and ask questions,' said MacGillivray. *Ignite, stand back.*

'Questions? How dare you? If you don't release me RIGHT NOW, there'll be bloody hell to pay. You'll be sorry you were born, MacGillivray.'

'It is routine, sir.'

MacGillivray's unruffled demeanour served to temper Pendle's furious indignation. He sat slowly back down on the bench and said, 'This is absurd, MacGillivray. Ask your bloody questions so we can all move on. Rome is burning out there, man.'

'Is what you said to Grey true?'

Pendle's hard stare held. 'What does he *claim* I said?'

'That the NSSA is systematically culling the nation of organised crime syndicates, and that it plans to roll out this covert culling for other sections of society, too.'

'It's common knowledge that we have emergency powers to act for the nation's security.'

'You admit it?'

Pendle ignored the question. 'Let me go *before it is too late,'* snarled Pendle.

'What do you mean by *too late*?'

Pendle relaxed; his fury eased by his veiled threat of retribution.

And, unfortunately, MacGillivray didn't think he's bluffing. *They were staring into a bleak future if it all went wrong.*

Pendle addressed Crozier, turning the knife. 'As for you, DI Crozier, are you with this *Deadman walking*? Or do you want a career when this debacle is over?'

*Had he taken a risk in trusting Crozier?* Thought MacGillivray. *Had his judgement been swayed by his moral duty to compensate for gender and diversity? What if she's a leak? Another plant like Grey? He'd only have himself to blame if he was wrong, and now, he'd find out the price for his naiveté.*

29

## DCI MacGillivray

'I believe you have questions that need to be answered, sir,' said Crozier. MacGillivray could almost hear Crozier scraping a line in the sand.

Pendle appeared not to believe his ears, judging by his following words. 'You bloody fool... you... you're finished, too.'

*How many people do the right thing when their grace is under pressure? Crozier had.*

Crozier turned to MacGillivray, making a faux surprised face and said, 'Who would have thought it? Someone who is who they claim to be around here!'

He laughed despite himself and felt slightly embarrassed that she had worked out he had doubts about her.

Pendle's furrowed brow signposted his confusion at their shared mirth, particularly in the wake of his dire threats. MacGillivray turned to Pendle, saying, 'Everything you are doing is wrong; people should at least be given a fair trial.'

'And waste the court's time and resources? And the prison budget? Only for them to be released to repeat offend. No, clear them out. I'd say we've got it right, even if we kill a few that may not deserve to die.

What's that cliched phrase? Oh, yes - *collateral damage*–that's the one.'

'Aren't you now a burden? Should we save the court's time and resources and save the prison budget...etc...etc...'. Crozier's words were mischievous, and she relished her effect on Pendle.

And MacGillivray couldn't deny he took some pleasure in his reaction, too.

'Now wait a fucking minute...! You're bluffing,' said Pendle, eyes widening.

'No. We won't do it. But we know someone that will,' said Crozier.

'Marsh? You mean you'd just hand me over to *Marsh?* That's barbaric!'

'He's big on irony, isn't he?' said Crozier.

'My task force will commandeer this place by the morning and you'll both be thrown in jail for treason.'

*MacGillivray could never countenance such acts against Pendle. Though Pendle isn't to know that. And he's yet to comprehend Crozier's ability to mess with people's heads.*

*And she had got something out of Pendle: the rough time of the NSSA's attack.*

'Come on, let's leave him be. For now,' said MacGillivray, employing his own brand of psychological pressure.

'You'll all be bloody sorry. You... you... *morons.*'

---

MACGILLIVRAY LED the way to his office.

'About the NSSA job - it offered a chance to save the station. Before all this culling came to light.'

'It's fine, I get that – and you not wholly trusting me, too.'

*She doesn't sound fine with it.*

'Look at this,' said MacGillivray, producing a map of the southeast of the country from his drawer. He rolled it out onto his desk; an outbreak of red dots spread across Kent and East Sussex.

'It's like a red army marching across the map,' said Crozier, interest piqued.

'You are not far off. The towns with red stickers have had high rates of *accidental* deaths in the last few months. But here's the thing: they've involved criminals, according to my contacts' reports.

'You got this info from the phone calls you made earlier?'

'Yes, from other police stations.'

'It's true, then. There's an attempt to cull the criminal classes. If the NSSA is this mystery X factor responsible for these deaths, then we can't trust anyone above us.'

'Yes, it appears we're on our own.' *With the closure of so many police stations, it often felt like he'd been holding a barren outpost. Now, it feels far worse.*

'We've got to get the story out there.'

'I have a friend who works at the BBC who might be prepared to help.'

'Can you trust him?'

'He's often bemoaning the state of the BBC and how it folds to the government's demands, complaining that the Corporation tries to walk an impartial path, but they avoid or water down reporting on sensitive issues for fear of upsetting the government.

I had better call him and see what he is prepared to do.'

---

'SCOTT? HOW ARE YOU?' asked Tim, delighted at seeing his friend but tinged with concern.

Time had crept by; MacGillivray's procrastination had been without peers, putting off all those awkward conversations about family, death and how he felt. Now, faced with the negligence of his friend, MacGillivray's guilt escalated.

'I've been busy at work. I'm sorry for not getting back to you.'

'You have nothing to apologise for,' Tim sounded as if he was at pains not to break the eggshells he's walking on.

MacGillivray was familiar with death in his job. But there was no inoculation for when it happened closer to home. Life had become about people missing–and he had shut his friends out. He hadn't just lost his family - he had lost everyone around him - *and That was his fault.*

Talking to Tim gave MacGillivray a glimpse back to his old self. He realised he had been avoiding this, of how the passing of time had shown him how he had changed before he'd become an emotional automaton.

'Is that MacGillivray?' said Tim's wife, Sally, in the background. 'When will we see him next?'

'Did you hear that, Scott?'

'I did. How about tonight?'

'Eh, that'll be great!'

'Though I must admit it's because I need your help, but I'm not sure what you'll make of it.'

'Well, you had better spit it out, DCI, and let me be the judge.'

'Can I tell you when I see you?'

'That's wise.' Tim immediately picked up on the possibility of eavesdroppers on the line.

. . .

'A SIXTY-MILE TRIP through the night, in this bloody weather? Hell, let's do it,' said Crozier. MacGillivray looked ruefully out of the window. *The rain's so dense as if it had done away with the need for air between its drops. He looked ruefully at the road overflowed with water and drains regurgitating dirty water.*

He packed the documents into a super seal case and placed it into his rucksack. 'I guess there isn't any point in ordering you to stay.'

'Only if you want an outbreak of insubordination,' she said with a ubiquitous glint in her eye.

'Somehow, that never seems far away with you, DI Crozier.'

'What about Holroyd and Grey?'

'We can't have them hanging around; they'll know we have left.'

'You don't trust them?'

'They may have changed sides, but I can't gamble on them when so much is at stake.'

'Agreed.'

THEY FOUND Grey and Holroyd sitting in the canteen, Grey looking wasted, struggling to hide his discomfort.

'Nothing is likely to happen until tomorrow. Why don't you get some rest? There's a hotel nearby, and they might even have hot food. Meet back here at seven am?'

Grey looked grateful for the chance to take time out, so Holroyd agreed.

## 30

## Holroyd

The Hawham Quality Hotel had seen better days, but now it had peeling wallpaper, chipped paint and well-trodden carpet. *Money was definitely an object here.*

*At least the sheets are clean,* thought Holroyd, once they had checked into their room - the last one left.

She inspected the two mugs on the sideboard, turning her nose up at the spider veins in them. 'Cracks; they harbour all kinds of germs,' she said. *It was a strange foible of hers to hate fractures in china so much, considering the life she had led.*

*Grey didn't seem to be bothered. In fact, he looked like he could kill for a good night's sleep–ironic, as it's probably the killing keeping him awake.*

Room service delivered a tray laden with two plates of chicken casserole and a bottle of wine. They set about their meals with indecent haste; both surprised at the good quality of the fare.

'You take the bed, I'll take the chair,' said Holroyd.

'We can both use the bed. I am sorry to disappoint you, but I won't be in the mood for anything romantic,' Grey said, wincing with his laugh.

'Here, stud, take these.' She handed him his

painkillers, antibiotics, and a glass of water. He ignored the water and took them down with his wine.

'You are so badass, right?'

Grey smirked, lay back against the bunched pillows and closed his eyes.

Holroyd cleaned Grey's weeping wounds, replacing the dressing over his strained stitches.

'Holroyd, I can't see a way out,' said Grey. She'd never heard him so downbeat before.

'When you've had some sleep, things will look better in the morning.' *Nursing and mentoring are not her style, but she had to say this.* 'Grey...'

Grey had apparently read her mind. 'Don't sweat it – you did what you had to, and we've been taken for mugs while trying to *execute our duties*.' Grey delivered the few words in an American drawl.

*How could he forgive so easily?* She was humbled, bemused. Rare feelings for her. 'Thanks. But whatever happens, promise me one thing,' she said.

'What's that?'

'Lay off the accents.'

31

# DCI MacGillivray

On the A24 going north, MacGillivray and Crozier passed the collapsed skeleton of the old hypermarket. It stood as an epitaph to an old way of life as the country shrunk back to regionalisation and artisan industries. MacGillivray didn't think that was necessarily a bad thing, which brought with it a reported reduction in pollution - if he believed the news reports.

The police pool car's engine rhythm melded into a white noise, only disturbed by the rattle from rifles and spare cartridges that lay on the back seat. The repetitive journey of the windscreen wipers worked at their maximum capacity, and still the rain could obscure their sight.

'Tell me something about Tim Salisbury,' asked Crozier.

*Is Tim a reasonable risk? That's what she's asking. And since she's putting her life on the line, so it's a fair question.*

'Tim's university thesis: *Are War Atrocities Increasing or Decreasing and Are the Rules of War Globally Observed?* had been taken up by the publisher *Overt House*, making it onto the Guardian's list of top political titles for the year. Tim was modest about it, but it was a big deal.

At the time of publication, Tim had also started his journo life in the local newspaper when Sally Myers, chief political editor at a renowned broadsheet, got in touch to invite him to lunch to discuss his book.

He told her he wanted to show the world it needed to be vigilant over its behaviours towards each other - especially in war. Sally said that anything that stopped *unnecessary* atrocities in war is a good thing. Her line in dark humour merely adding to the appeal of Sally for Tim.

Sally offered Tim a job. And, after a month of agonising over work lines that shouldn't be crossed. He asked her for a date.' MacGillivray smiled at remembering Tim's acute agony. And at his relief and joy when Sally accepted his proposal. 'You took your time risking a harassment case,' she'd said wickedly.

Crozier exuded a warm chuckle. 'I like the sound of her.'

'The Broadsheet had been a habitual thorn in the government's side. So, it could not rely on government bailouts when times became difficult. The business fell into a death spiral, leaving Tim and Sally out of work. The BBC threw Tim a lifeline and offered him a job as a political commentator. But, in increasingly fraught editorial meetings, Tim questioned the government's influence on what the BBC reported until Tim's manager took him aside and told him to *toe the line or leave*. It wasn't what he'd been used to in his last role. Tim decided it would be better to be inside the system to instigate change when the opportunity arose, than to be on the dole. He once confided in me that he doubted the moment would ever come.'

'Until he got the call from you.'

'Until he got the call from me.'

'It sounds like he's a good person.'

'Yes, I think so. We're running on fumes; we need fuel,' said MacGillivray.

THE PETROL STATION'S sign stood as a beacon in the rain, guiding them in. The forecourt is an oasis of tired, dry concrete streaked with rain streaks from the leaking roof.

MacGillivray parked next to the first pump, switching off the engine.

Another car, an old Audi, was parked by another pump.

A man stood next to the garage shop, urinating unselfconsciously against the garage shop wall.

'Wildplassen,' said Crozier, disgusted. *Dutch for urinating in a public place,* he remembered.

He's joined by another male leaving the building with a basket full of snacks and beers, one open in the other hand. The intention to settle his bill missing from his to-do list, judging by the shout of *Hey, you need to pay*, from the shop assistant.

'You've forgotten my Fosters, you dick,' complained the pissing companion.

'Help yourself, then.'

*Should I let it go?* MacGillivray asked himself.

Crozier coiled, waiting for his word.

A middle-aged male attendant, wearing a bright, home-knitted yellow wool cardigan and a matching hat, ran out of the forecourt shop. He acquiesced at seeing the sudden appearance of the thieves' knives, opting to return to the store.

'Good decision, asshole,' said one of the nonpaying patrons as they followed him back in to steal their beverages of choice.

THEY RETURNED TO THEIR AUDI. The driver, confused at not finding his bunch of keys in the coffee cup well, shouted to his mate, 'Where're my fucking keys?'

'Pockets?' suggested his passenger.

MacGillivray whistled, catching their attention, the keys hanging from his index finger. 'You left these in your vehicle. Careless; they could have been stolen.'

They stared at MacGillivray. Then, the previously urinating one said, 'Hand them over.'

'Lay the knives on the ground, pay for your purchases. And we'll let you go.'

They processed this, their petrol station raiding high spirits, stalling.

'*Let us go?*' he said incredulously. 'What's to stop us from coming over and cutting you up?' *The driver sounded educated, university level. The institution would be disappointed to learn that its cultural ethos had been interpreted as Rob Everyone Blind. Or perhaps not, judging by the escalating fees of the educational institutions.*

'Nothing. Except maybe me throwing your keys into the trees?'

'Yes, but it would be worth it.'

'Eh, that's not a good idea. We'll be stranded out here,' advised his wiser companion.

The less wise man with a knife took a step forward. His ego and alcohol were melding into an easy alliance.

Crozier held up her rifle and clicked off the safety.

Two sets of thieving eyes tracked from MacGillivray to Crozier.

'Last chance,' said MacGillivray.

The one with a healthier sense of self-preservation dropped his knife, while the other kept coming.

Crozier sent a shot into the ground ahead of the perp. The garage attendant dived to the floor, presumably expecting the fuel tanks to explode. He'd need to have a *Health and safety* chat with Crozier later.

The man laid his knife carefully on the ground. 'Okay, okay,' he said. 'Why do you care, anyway? You're all going to be out of work soon.' *No wonder they think they can get away with anything; the rumours are already spreading about the station closing.*

'Go. Before I change my mind,' said MacGillivray.

They paid and went.

'THANKS. IT'S THE FRIGGIN' Wild West out here these days,' said the cashier.

'I've got their registration; we'll check them out later,' said MacGillivray.

'The fuel is on us. Here's a little something.'

The service station attendant gave them each a bar of dark chocolate–with the shortage of cocoa beans, it is a luxury item.

Sitting in his car, MacGillivray, still buzzing from the incident, crammed a line of chocolate into his mouth. He barely allowed the rectangles to warm up, fracturing them into cold shards, savouring the cocoa and sugar hit.

Crozier let out a tension-busting laugh. 'Good, isn't it?'

He nodded, 'Yes, let's crack on. We've lost enough time.'

'But it was good to make a difference, right? To stop real crime?'

'It was.' *It felt pure.*

## 32

# DCI MacGillivray

'We've got company', said MacGillivray, peering in his rear-view mirror. In the distance, bright lights reflect off the rain-slicked livery of the NSSA vehicle.

'How'd they get here so fast? Someone must have betrayed us,' said Crozier.

'It's a little academic at the moment.'

Something metallic bounced and whined off MacGillivray's car.

'They're *shooting* at us!' said Crozier, startled. 'They're going to catch us at this rate, pull over.'

MacGillivray pulled into the breakdown lane without question.

'What are you going to do?'

'Watch.' Crozier grabbed a rifle from the back seat and scrambled out. MacGillivray got out, rewarded by great water droplets falling from the trees' stark cover. He watched her lean on the car roof and flex her trigger finger, preparing her sight. Crozier's first shot hit the wheel cover. Her second and third hit tyres, deflating them, causing the lorry to drop lopsidedly, sliding to the side of the road. A roadside metal barrier buckles, gamely trying to hold the truck back. It comes to rest, tilting over a drop into a small gulley.

Crozier punched the air. 'Yes! Get in there!'

'Good shooting, Crozier. That'll buy us a little time.'

'Wait... they're unloading something.'

MacGillivray squinted across. 'Actually, something is unloading itself.'

'Godverdomme, it's a *robot,*' said Crozier.

MACGILLIVRAY TRIED to peer past the rear blue light fanning dementedly from the lorry's rear. He took in the rolled steel and the large sensory eye, glowering in the greyness. The machine's circular stainless-steel body is mounted on three black traction balls, offset by two metallic extruding tubular arms. Its body is topped off with a metallic ball of a head, its sleek sensory panel seeking, if the police briefings were to be relied upon, heat signatures. *Theirs.*

The robot stopped rotating, pointing towards Crozier and MacGillivray, like a Doberman Pinscher picking up the scent of its prey. Its focus sending a fearful chill down MacGillivray's neck.

'Let's go,' said MacGillivray, trying to keep the nerves out of his voice.

He drove them away, running the gauntlet of shots from their marooned, would-be assassins.

'That's a Peel Bot,' said Crozier as they left it in the rain-smudged distance. 'You said they were only used for city curfew duties.'

'We should feel flattered they're using it on us. The machine's a prototype, with a limit to the speed it can travel.'

'Let's hope we can outrun it, then. I can't stand machines, I really can't, MacGillivray.'

*Crozier has technophobia.* It's odd to be afraid of something pre-programmed. Now, being chased by a machine, it seems a different proposition. Particularly as it's staying in his mirror. There was something darkly terrifying about a machine that could hunt you down–perhaps more so than something born of flesh and blood.

'They're supposed to travel at a maximum of six miles per hour. They appear to be getting a lot faster. And I have more bad news - we've got a fuel leak,' said MacGillivray. The car's gauge taunts them with its alarmingly fast descent towards empty. *Each extra mile will be a precious gift.*

'Shit,' said Crozier.

MacGillivray ignored Crozier's frustration, concentrating on eking out the remaining fuel by crawling along and increasing their distance from the robot.

Then, the car engine's tone went off a cliff just as they made the brow of a hill and cruised a little further before coming to a halt.

They clambered out, MacGillivray's ears humming with the ghost of the engine's noise in the enhanced quiet.

Crozier said, 'Here look: a bullet hole–they got lucky and hit the lower side of the fuel tank.'

'We got lucky that it didn't ignite,' said MacGillivray. 'Come on, I know this area; it's only a short distance.' MacGillivray grabbed his rucksack containing the precious cargo of proof, along with guns and spare clips.

They were near a populated area. But out here in

the woods, in the heavy rainfall, they were at the mercy of a hostile environment.

The orange tinge of the town light is missing. Light is power, and power is scarce, so it's switched off; everywhere is succumbing to the dark of night.

MacGillivray spots the subtle blue light from the Peel Bot cutting through the rain haze behind them. The rays of indigo randomly picking its way through the naked trees, closing in on them.

MACGILLIVRAY IS BACK on familiar ground. Having gone for walks with Tim through these woods on their way to one of their favourite pubs. He remembers a small river they could cross and maybe even lose the robot. 'I know a shortcut,' he said, leading Crozier towards a wooded hill.

CLOSE TO THE TOP, MacGillivray is suffering. His exhaustion insidiously blunting his mind, making him lose connection with his legs. He trips over a tree root, face-planting into wet forest mulch, cold sliminess and twigs stinging his overheated face.

Detective Inspector Crozier retraces her steps and pulls at his arm, coaxing him to get up. 'Move, sir,' she said.

*She rarely called him sir. We must be in serious trouble.*

He implored his aching muscles to work. His breathing rasped, as if hot sand lined his throat. While he tried to summon energy, the metallic servo sounds of their pursuer came nearer, escalating his terror. His world had compressed into two illuminating choices.

Option 1: keep running.

Option 2: or die.

MacGillivray clamped his teeth and moved, choosing option 1. They reached the brow of the hill, the sound of the river softened by the density of rain-sodden trees.

MacGillivray looked back, checking for the machine's progress. He tries to hear but catches little through his laboured breathing and his blood pounding in his ears.

MacGillivray's survival instinct screeched inside him, an irate chimp panicking: *What if the river's too deep to cross?*

*They would find out soon enough.*

Then, the stretch of water came into view. It was deeper than he thought it would be and is moving quicker, too.

'WE'RE GOING IN *THERE*?' asked Crozier, less a question, more a statement of trepidation.

'If we cross the river, we can leave the robot behind.'

'Godverdomme. Let's get on with it.'

MacGillivray glanced behind and saw their pursuer breaking from the green darkness, splintering branches, spraying water from foliage in their hunter's wake. *Did a blue light ever look so deadly?*

Despite the thickening mud, Crozier led surefootedly as they came out from under the tree's canopy, approaching the water's edge.

MacGillivray is dying on his feet, blowing hard but not ready to give in.

Crozier stopped, sighted her rifle and pulled the trigger.

The Robot's well-synchronised arm shot up; a metal guard plate attached to it deflecting the bullet.

She fired again. Nothing exited the weapon's chamber. It's empty. She discarded the weapon. 'Come on,' said Crozier. She slid down the bank. MacGillivray followed. Plunging into the chilly embrace of the black water—shock forcing him to take a sharp breath.

The strident tug of the current pulled at him as he struggled to stand on the shifting riverbed of gravel and mud. His clothes and rucksack were now heavy with the weight of water, the torrent threatening to drag him down.

Crozier negotiated the water back to him and shouted to be heard above the wash. 'If you get swept away, do not get caught under the trees. You won't get out.'

MacGillivray nodded; freezing to death or drowning would be the likely outcome of that scenario.

'Take hold of this,' she said, offering him the knotted end of a thick branch. Crozier could keep her distance from him, so if the swell dragged him away, he wouldn't grab her if he fell. *Smart.*

The Peel Bot picked its way down the path to the river, traction wheels disturbingly efficient.

They made for the other side. MacGillivray could soon feel numbness in his extremities. *I am losing control of my limbs.* They were near the mid-point of the crossing, bobbing on toes, trying to avoid rushing guillotines of sheared wood and debris. *It's like some sadistic old TV game.*

Crozier pulled on the branch, trying to lead him, but he stopped progressing; just keeping his footing is taking all his energy.

Timber flotsam struck MacGillivray on the side of his head. He felt dizzy. Water washed over his head, entering his mouth, shocking and jolting him back to clarity.

'MacGillivray! Are you alright?' shouted Crozier over the gush of water.

Lack of sleep, food, work overload, months of emotional turmoil, and little exercise had not prepared him for this; he's at the limits of his endurance, the deadly weight of despair forcing him down. *Death suddenly seems like a sweet solution. The one thing that could take away his pain. It would be easy to give in. To go under. But would it be a fitting tribute to his wife and child? And he desperately wanted Crozier to live, too. And what of the town? It's still possible to live for others. To offer something greater than the sum of his pale parts. To find purpose above his sorrow.*

He felt a resurgence of will and anger, displacing his despair, and he struck out for the far shore with newfound energy reserves.

The tug of the undertow is pulling at Crozier's legs, severing her hold of the umbilical cord of timber that had joined MacGillivray to her. He grabs for her, but she's gone, floating downriver.

'Crozier!' he shouted.

He could feel the perfidy of the riverbed, the gravel dislodging him, but still, he held good, fighting for breath as the river washed over his head. To his amazement, the bank's closer than he thought, and he is now under the canopy of overhanging trees.

Then he sensed the whiplash of something flying across the river and heard the metallic clunk of something hitting wood. *The robot had shot a grey line above him, tethering it to a tree stump.*

Through the noise of the river's wash, MacGillivray could hear an engine whirring.

With icy dread, he turned and saw the Bot pulling itself, skimming across the water. Its ball-shaped tri-tracks acting as buoyancy aids; the machine's weight held the wire taunt.

MacGillivray reaches with his branch and snags a grip on the wire, pulling himself along. *The water only reaches his thighs - he can get to the bank.*

Exhausted, reduced to his knees, he crawls out, legs crunching on stone. Coughing up water, he cannot get oxygen fast enough into his body. The Peel Bot's making slow, steady progress along the zip wire. One thought beats a path to his numbed mind: *disengage the zip wire.*

Using the wood that saved his life, he wills it to do so again, using it to hammer at the claw that binds the line to the tree. The Peel Bot is now two-thirds across the divide, navigating through floating timber.

But at last, its luck runs out.

A tree has snagged the zip line behind the Bot, its branches tugging, slowly bending the wire downstream. The branches splinter, shoot up, but the tree keeps its stubborn grip. Tree, river, and wire are locked in a fight for supremacy.

The Bot is trying to realign itself, inching towards the bank.

MacGillivray, spellbound by the Bot's progress, breaks his trance and strains to release the line. The branch fractures and he can barely feel the damage to his cold, ravaged hands, noting blood from a gash wound to his left hand.

'Come on, let go.'

The tree's bark turns a sickening blue as the robot draws nearer.

Then, the miracle happens. MacGillivray can see the slow reveal of stripped bark on the tree as the hook relinquishes its grip. The line flailing back into the river–*and the robot is gone!*

That damned machine is careering down river, hammering into tree debris, the river's force trapping it into banked green spoil.

MacGillivray finds the energy to stand and punches the air, despite how heavy his arm feels.

But the question of what has become of Crozier soon quells his joy. He pans around and spots Crozier lying on the stony bank halfway between him and the Bot. She's hardly moving.

Then, the Peel Bot's tubular arms snap wooden branches, extricating itself with a dreadful quickness, working its way around the trapped wood to the shore. *Towards Crozier.*

MACGILLIVRAY HOBBLES as fast as he can, shouting and pointing, trying to warn Crozier.

Crozier lifts her head and sees the bot moving inexorably towards the shore's bank

and her.

Crozier screams at herself to get up, raises herself to her feet and moves. She is limping, grimacing, determined.

MacGillivray takes off his rucksack as he runs, remembering he has a handgun in it.

Crozier stumbled as she tries to quicken her pace.

*The machine is closing the distance on her. It's going to reach her before he does.*

Stumbling, she reaches for a small rock, turns and swings for the Bot's extended arm. Its steel grippers latch onto the edge of her coat. She launches a kick with her other leg at the Bot's head.

Its other gripper holds a needleless syringe, using that arm to fend off Crozier's kicking leg.

*Maybe an induced heart attack is easier to cover up than a broken skull,* thought MacGillivray.

He points his gun at the Robot and risks hitting Crozier. He pulls the trigger - *nothing: the water has ruined the weapon.* He runs.

An age passed as he covered the ground between them. He smashes the handgun against the Bot's syringe capsule, breaking it off, dark liquid running out of its glass and metal wound; the ruined capsule ejected onto stones, like a discharged gun cartridge. To MacGillivray's dismay, a new one slides into place from a hidden cache in its upper arm.

'Fuck,' moaned Crozier. 'Run, MacGillivray.'

'No chance.'

MacGillivray swings the gun again. This time into the sensory band of the Robot, rewarded with plastic cracking. The Bot lurches, releases Crozier's coat. Its sensors, perhaps damaged, struggling to filter now scrambled data.

MacGillivray swings once more, only the Bot dodges his strike, swinging at him with a counterattack, causing MacGillivray to fly meters away.

He lay stunned, his shoulder pulsing from the impact.

*Crozier!*

Crozier grabs a plastic shard of broken robot face plate and drives it into the Bot's vulnerable wiring.

A ten-second countdown in red characters appears on the damaged Bot's display, pulsing on each gauged beat. A metallic neutral voice proclaiming the Peel Bot is compromised and that it will self-destruct.

The malfunctioning Bot holds up its syringe, shaking erratically, about to plunge it into Crozier, its last malign act.

Only Crozier lunges with a rock, striking at the clutching arm of the Bot, rendering its gripper digits fractured at the joints. She scrambles free, the syringe swiping into fresh air.

Crozier staggers towards MacGillivray, helping him to get to cover behind a tree.

The Zero now reached on its countdown display, and the Peel Bot erupts in a concussive ball of flame. MacGillivray can feel ripples of heat emanating past their wood shield. The edges of the tree's bark catching alight, crisping.

33

# DCI MacGillivray

They walked into the deserted village, its chocolate box aesthetics only serving to underline the surreal nature of what they were doing.

In the distance behind them, they could catch the smoke from the Peel Bot's fire.

'It seems a little extreme for the NSSA to have the Peel Bot programmed to explode,' said Crozier.

'I think the NSSA are determined to give as little away as possible about their modus operandi, not least how their bots work.'

'That makes sense.'

'HERE WE ARE. We've made it.'

The semi-detached house evoking memories of how it had been when he visited them with his family - that other land existing only in his memory.

But Tim's place now offered hope–*and isn't that a kind of progress?*

Tim opened his front door, backlit by the light from a welcoming interior, illuminating the plush thatch of his black hair, his goatee beard sporting a few flecks of grey.

Tim is happy to see MacGillivray but can't hide the look he gives–the one that says, *fuck, you've aged,* before remembering himself and hugging MacGillivray.

Tim's wife, Sally, with her auburn hair, smooth-fitting yellow jeans, and a lime green jumper, brushes past her husband before attempting to squeeze the life out of MacGillivray.

'Are you going to introduce us to your friend?' asked Sally, eyeing the woman behind him.

'This is DI Lerato Crozier,' said MacGillivray.

'Lerato. That's a beautiful name.'

'It means *love* in South Africa. My mother's family lived there before migrating to the Netherlands. She wanted me to have a link with the old country.'

'I'm Sally. It means to charge out from a besieged place against the enemy.'

'Appropriate, considering what we're asking you to do,' said Crozier.

'It's raining, but you're both *seriously* soaked through,' said Tim, observing puddles forming on their tiled floor.

'That'll be from crossing the river by the wettest way: through it,' said MacGillivray, adding, 'A Peel Bot was hunting us.'

'Bloody hell! What happened to the bot?' said Tim, incredulously.

'It exploded.'

'Exploding Robots? What *have* you got yourselves into?'

'Before you reveal all,' Sally said, 'I'll dig out some dry clothes before you both die of pneumonia. Get the kettle on, Tim, and can you bring those cakes out of the oven?' instructed Sally.

'Yes, of course.'

. . .

'SO, what've you got for us?' asked Tim.

MacGillivray opened his rucksack, removed the case, popped its seals, and passed the files to him.

Tim skim-reads the documents, passing pages onto Sally.

'Well?' asked MacGillivray.

'It's one pit of shit you have got yourselves into,' said Tim.

'It's good to know Tim has lost none of his journalistic turn of phrase, isn't it?' said Sally. 'I hate to say this, but by just reading this, we're now in this *pit of shit,* too. This is *hot.* It could bring the NSSA down. Maybe even the government,' she said, MacGillivray's material in her hands, striding agitatedly back and forth in front of the open fire. '

Tim nodded his agreement. 'I had to sign an NDA to work for the BBC. I'm going to be breaching my contract.' *It sounded like Tim was stating a fact rather than his intent,* thought MacGillivray, but his stomach clenched anyway.

'This is dangerous. I wouldn't blame you for not believing it or not wanting to be involved.'

'It's believable, alright, going by the government's attempts to impose their influence on the tone of our news output. They've got the BBC in a vice grip, using the threat of litigation and the withdrawal of budget contribution. If the BBC folds, it could be the end of mainstream journalism as we know it. And haven't people become desensitised enough to what the government and corporations are doing? The government is using that apathy to its advantage, keeping the power and wealth with the few. If we do nothing about it, we won't have any freedoms left at this rate.

But it's become harder to see what we can do. So, if you have a plan, I want in.'

'I've been battering Tim with criticism of the Government, the NSSA and the BBC ever since he joined. We must be involved,' added Sally.

MacGillivray let his breath out, relieved at being proved right, that the *railing at the gates of injustice,* Tim and Sally, are in.

'Okay, great,' said MacGillivray. *Hope lives.*

'Right now, you're our best hope for saving the country. I can't sugar-coat it.'

'I can't deny we're feeling the pressure,' said Tim, reaching for Sally's hand.

'It looks like sleep, and our precious stash of coffee, will be sacrificed in shaping the script tonight,' said Tim, rolling his sleeves up in eager anticipation.

'When can you release it?' asked MacGillivray.

'Eight a.m.'s a good time to catch listeners. And, with some luck, I'll get this out before anyone can stop me.'

'Good,' said MacGillivray, the strain easing from his face. 'It's been good to see you. Next time we see each other, we can celebrate our success, but we have to get back: we've got Pendle in custody,'

'*The* Pendle? The head of the NSSA?' said Sally, flabbergasted.

'Yes. The NSSA could storm the station in the morning to free him.'

'Hell's fucking bells,' said Tim.

'We've got a couple of issues. Our transport took a hole in the tank and a flat tyre. We can solve the tyre with a sealant. But we're stranded without fuel.'

'Take our car,' offered Sally.

'We can't risk it; they'll trace your car back to you. Then they'll figure out what we're doing.'

'That's a good thought,' said Tim.

'If you could plug the tank, would that help?' asked Sally as she rushed off to the kitchen, returning carrying a tapering metal champagne plug with rubber seals and a hammer.

'It might work,' said Crozier, approvingly.

'I'm going to syphon fuel from our car for you,' Tim said as he left.

A few minutes later, Tim came back with a tank of fuel. 'Let me at least drop you off.'

---

DRIVING BACK in MacGillivray's car, a sign advises that they are two miles from Hawham. Through the early gloom, MacGillivray can glimpse the NSSA road-block. He switches off his lights, slows and parks in a layby. 'We're on foot again.'

'At least it's stopped raining,' said Crozier.

They crossed a couple of fields, progress slow in the dark, the draw of the silhouette of town guiding them in.

They came out onto a minor road, the rumble of lorries softened by mud and silt on the flooded roads, causing them to run and hide behind a convenient wall. There are large letters of *NSSA* on a passing can-vassed lorry.

'They mean business,' said Crozier.

'Let's reach the station. If we keep Pendle at bay, we can buy some time until Tim's bulletin.'

A female voice boomed through a megaphone.

'There is a curfew in place... stay at home... this is for your safety.'

MacGillivray and Crozier edge across the road

after the last vehicle had passed. The police station is now in sight.

Then, two black-clad, rifle-wielding NSSA operatives on foot glanced their way, snapping into action; one shouted for them to stay where they are. Then, seeing MacGillivray and Crozier break into a run, MacGillivray sees one of the NSSA operatives raise their rifle and fire.

## 34

# Holroyd

Grey had gone out like a light last night. *His physical scars would be pronounced, the emotional ones underneath his 'don't fuck with me' face, less so,* she had thought. Holroyd felt even more confused about what she felt for him. *She had needed rest, not more conundrums.* But it had been a misplaced hope as the night brought her little sleep. Holroyd had killed drug dealers, human traffickers, pimps, assassins, some of the most vicious bastards walking the earth and hadn't lost sleep over their passing. But being manipulated into killing for a government to stay in power enraged her. She should have seen through the government's and Pendle's, deception. She isn't naïve, but she now has an unusual crisis of confidence in her judgement.

*And now that they were on the run, they would never be safe. So, they'd have to make a stand and hit back while they could.*

NOW, early this morning, noisy, heavy engines ominously passed the hotel.

Already dressed, she went to investigate, discreetly hiding by the trees near the entrance door.

Two dark blue Jenkel Guardian NSSA vehicles moved slowly again, passing with aggressively angular bodies and an uncompromising rumble.

Trying to remain as unobtrusive as possible, she followed the NSSA wagons, watched them stop and observed personnel crisply disembark and unload materials. *Not good news.*

The road is now sealed tighter than a fresh head gasket, with intimidating barriers, intimidating NSSA operatives wielding intimidating rifles.

Some early-to-rise townies, trying to get past the barrier, are politely turned away. They raise their voices in protest, anger turning to fear as the initial NSSA politeness sours into assertive posturing. Their weapon's safety catches coming off in a snapping sound, leaving no room for negotiation, as they drive civilians back.

*Things are going bad here fast.*

Hurraying back to the hotel, Holroyd is surprised to see MacGillivray and Crozier crossing the road. *They looked dead on their feet. What had they been doing?* Then she saw the two fast-approaching, murderously purposeful-looking NSSA operatives.

And they are shooting at the detectives.

## 35

# DCI MacGillivray

Bullets fly past MacGillivray and Crozier.

*They'd been caught in the open; anywhere safe is too far away to run to. This is it.*

Crozier was faster than him, so he ran behind her to give her some cover.

Then contradictory flashes of rapid-fire came across the road, and he and Crozier held their arms to protect their heads. But the sounds were followed by the jerking fall of the two shot NSSA gunmen, emitting agonised cries of pain, and then they fell coldly silent.

MacGillivray froze and stared. *He is relieved and surprised to be alive.*

'For fuck's sake, snap out of it and get to the station. I'll get Grey and we'll meet you there,' shouted Holroyd, their potty-mouthed saviour.

---

THE POLICE STATION came into view, *their sanctuary, at least for the moment.* MacGillivray endured a laboured climb of the stairs to the entrance, again following Crozier. The early morning darkness is thankfully all-encompassing on the naked steps when he expects bullets to come for him at any moment. Crozier punched in the six-digit code to open the door and got it wrong. MacGillivray laid a steadying hand on Crozier's shoulder. 'I've got this, thank you,' she said spikily, yet her eyes relayed an undertone of gratitude. She tried again, successfully. They slipped into the building, relieved to feel its dry warmth.

Neither Jacobs nor Hanbury are at the front desk, adding to the Marie Celeste feel of the place, feeding MacGillivray's unease.

'What's happened now?' he asked.

Stiff-legged, he moved as quickly as he could down to the cells, receiving a jolt when he sees Hanbury, bound and gagged on Pendle's cell floor. *There's no sign of Pendle, and Sam Marsh is missing, too—there's been a complete jailbreak.*

'Hanbury's hurt,' he shouted.

He rushed to get the keys to Hanbury's cell, got in and eased the gag off him.

Crozier appeared with a med pack. She crouched down next to Hanbury. 'You've got a nasty gash on the back of your head; you could have a concussion.'

'I think Jacobs attacked me from behind,' said Hanbury, reviving.

'Easy Hanbury, we'll deal with Jacobs later. How are you feeling?' asked MacGillivray.

'Groggy, sir. Nothing a month's holiday on a warm beach won't cure.'

'There's nothing wrong with your sense of humour, at least. But let's get you down to the hospital and get you checked out. But do you know where Pendle went?'

He shook his head slowly, grimacing at the movement. 'No, sorry.'

The clock on the wall showed 7:20 am - *forty minutes before Tim's hoped-for rogue news bulletin.*

Then the entrance intercom buzzed. MacGillivray viewed the camera monitor, seeing Holroyd and Grey, then let them in.

'YOU'VE HAD A CLOSE SHAVE, I hear,' said Grey.

'So, what have you been doing?' asked Holroyd, eyes lasering into MacGillivray.

'We passed evidence onto someone that might be able to release it into the public domain, I'll tell you that much,' said MacGillivray.

'It can only be radio,' said Holroyd, thoughtfully chewing on her lower lip. 'You mean the BBC, don't you? You're going to broadcast the information. A clever move, if you can pull it off.'

*She's quick*, thought MacGillivray, sharing a startled look with Crozier. Even though she'd saved their lives, he still could not trust her to confirm what she said.

Instead, he said, 'I have bad news: Pendle escaped.'

'For fuck's sake, couldn't you have at least trusted us to guard him while you were away doing what you were doing?' asked Holroyd.

'Would you trust us in his position?' Grey asked magnanimously. Holroyd shook her head in resigned disgust.

'We think they will be raiding us soon, but do you have any idea what else the NSSA are planning?' MacGillivray asked.

Holroyd looked at Grey, appearing to agree to something.

'Have you heard of the Town Hall fire in Bristol?' asked Holroyd.

'That's a few months ago. It was a terrible, tragic accident. There were so many people killed,' said Crozier.

'It wasn't an accident,' said Grey flatly.

'They, the NSSA, were covering up a leak of Project Deadhead?' Crozier.

Holroyd and Grey nodded.

Grey continued, 'One of our operatives had fallen for a local council worker, sharing pillow talk about Project Deadhead with her. He'd followed his dick, then his heart, the bloody fool. She informed the Head of her department, who called the police. They instructed the council worker to help them trap the operative. Only wracked with remorse, the council worker broke down, confessed to our agent about what she'd done, and he got away by the skin of his teeth.'

'The operative contacted Pendle to admit his mistake and ask for help. Pendle sent an NSSA team in.'

'Couldn't anyone phone out?' asked Crozier.

'The NSSA commandeered the telephone exchange. They *interviewed* them and gained the names of people they had told. They rounded all those up and escorted them to the Town Hall.'

'Then the NSSA killed them?' asked Crozier.

'Yes.'

'How the hell can you live with this?' asked Crozier.

'I am not sure,' said Grey, wearing resignation.

MacGillivray had heard rumours about Bristol. *The conspiracy theory enthusiasts had been ridiculed.*

*But they'd been right after all. The government had been prepared to let the bodies pile up in the name of their power-grabbing agenda. Even now, in his naiveté, he struggled to comprehend that his own UK government would do this to their people–but here is irrefutable testimony.*

MacGillivray burned with anger. *They had to be stopped. And his hopes rested on his friend Tim.*

'To answer your question, they're going to do something far worse–they are going to wipe Hawham off the map. They can do that. They'll report that another virus has been found and then partition off the town. Then they can claim that the people had all died in quarantine–but really, they will cleanse the place of any knowledge leaks–and that means all of us.

So, let's hope your plan works,' said Holroyd. She ran a hand through her hair and said, 'It's settled: we wait for the news release. I need coffee. Anyone else?'

*It was such a prosaic thing for a killer to do: to put the kettle on.*

---

MACGILLIVRAY WATCHED the second hand on the wall clock trudge the last interminable circuit towards eight o'clock.

The BBC's radio station played Montague and Capulets, a dark, dissonant piece that MacGillivray thought was in keeping with the mood.

Tim's voice came on the radio, asking for the Nation's attention for an important announcement.

'Come on!' shouted Crozier, jaw strained, fists clenching, eyes alive with hope.

Then gunfire erupted over the airways, mixed

with the splintering of wood and shouting, dominating the broadcast.

MacGillivray held his breath, waiting for his friend to deliver his missive despite knowing in his heart that it would never come.

*His heart was proved right. Nothing. A void of silence stretching achingly out.*

MACGILLIVRAY PICKED up the radio and did what many non-technicians did: he banged it.

The radio declined the invitation to respond to abuse.

After fifteen long seconds, the radio erupted; the sound of the old classic *No News is Good News* from *Newfound Glory*, making MacGillivray jump and drop the radio.

'They've used music to fill the dead time,' said Crozier, shuddering at her unfortunate choice of words. She threw her mug, shattering it against the wall. 'So. Fucking. Close. If I ever catch that bastard Jacobs, I will snap his neck,' she said.

Despite her savage, righteous anger, she slumped back into her chair, utterly deflated.

MacGillivray hunted frantically around his brain for an alternative solution.

*Desperate to avoid the one choice he had left to him.*

*And that was to ask the killer of his family for help, because only Sam Marsh had the resources to carry the fight.*

*It was fate at its most ironic, cruel, worst.*

# 36

## DCI MacGillivray

'What now? I'm out of ideas.' Crozier's voice sounding filed down with tiredness and despair.

It hurt MacGillivray to see Crozier in this state, which made it a little easier to focus on what he had to do: to ask the murderer, Sam Marsh, for help. But he still struggled with inner turmoil, trying to keep his want of justice beneath the town's needs.

'I need to get out to Marsh's place,' said MacGillivray between gritted teeth.

'Why?' asked Grey.

'He might help.' Even now, the reflex was reluctance to divulge more information to him.

'How can you morally justify collaborating with a criminal?' asked Grey.

MacGillivray raised his eyebrows incredulously and said, 'I'm working with you and Holroyd, aren't I?'

Crozier stood next to him; he could feel her *solidarity*. 'Marsh *is* the only one who can help, isn't he?' she asked with a tone that understood the cost involved for MacGillivray.

'But he's only just broken out of jail; will he even be there?' asked Holroyd.

'Unless you got a better option, what choice do we have?' MacGillivray's words met a stumped silence.

'I'll take you,' offered Holroyd.

'Then you both need to get away before your NSSA ex-colleagues arrive,' said Crozier. 'When they do come, we'll keep them distracted.'

Holroyd held her gun up and smiled chillingly enough to bring the temperature down further. 'Excellent,' she said.

Then, they heard the portentous sound of a heavy diesel engine approaching the police station.

Holroyd rushed to the window. 'They're here,' she said flatly. 'You'd better get on with your *distracting*.'

---

THE ARC of an NSSA lorry's Cyclops white beam hit the police building courtyard, accompanied by the ominous grinding of its gear cogs.

MacGillivray should be used to gunfire but is still shaken by the violent recoil of Holroyd's gun.

One of Holroyd's shots hit the driver, causing him to slump. *Another death–how many more will there be?* Thought MacGillivray. The lorry bumped to a halt against a column of the police courtyard entrance. The stone cracked, and one of the blue lamps that had sat there for over a hundred years tumbled, causing its glass, within its forged metal frame, to fracture.

A dark-blue-clad NSSA operative fired at them as he got out of the lorry's cab. She fell prey to Grey's headshot, a spray of a fine red mist issuing forth from her cheek and the exit wound at the back of her head.

MacGillivray fired his handgun, keeping the attackers from shooting at them. Crozier shouted at

MacGillivray to get his attention, tossing him a spare cartridge of bullets. She moved, crouching towards the staircase.

'Crozier! What are you doing?'

'I'm going to cover your escape,' she shouted as she hit the first rises that took her quickly out of sight.

Holroyd ran down the entrance steps, took flight off the brick steps and landed efficiently, taking cover behind a police Land Rover. She fired the last of her bullets, then smoothly changed the cartridge casing for a fresh clip of bullets.

NSSA operatives clambered out of the lorry and lined the car park's three-feet-high wall, working weaponry hard, creating a soundtrack of breaking brick, disintegrating wood and surrendering glass. Then they unloaded something from the lorry. Something nasty. *They've got a bloody bazooka,* thought MacGillivray.

A hastily launched bazooka missile screamed for attention, aimed at Holroyd. It hit the police Land Rover, causing it to lurch, catching alight. Then the fuel tank joined the fire party, its explosion adding fiery presence, its metal skeleton semi-exposed.

Flames leapt from the stricken Land Rover, blackening police station brickwork. MacGillivray's stomach churned with the smell of burnt rubber, melting metal, diesel - and a concern for Holroyd. *What had become of her?* Then, she appeared by him, like a genie from the smoke.

'Were you missing me?' she said, giving him a wink.

MacGillivray turned to see the bazooka operator reloading; he raised his gun, fired and missed. The Bazooka man pointed his weapon straight at MacGillivray.

A shot rang out from above from Crozier. The Bazooka man fell to his knees, spinning sideways with an arm injured, his released missile sailing wide to hit a tree. An NSSA colleague retrieved the dropped weapon, pushing another shell into the Bazooka's tube, preparing to fire at Crozier's window.

'Get out, Crozier,' shouted MacGillivray.

Upstairs erupted with the impact of the shell, and the roof breached. Masonry, dust, and debris fell, littering downstairs as the ceiling collapsed and the fire took a hungry hold.

'Crozier!' MacGillivray cried.

## 37

# DCI MacGillivray

*What? Crozier gone?* MacGillivray felt he couldn't feel more pain, but he's proven wrong. Fresh waves of anguish washed over him for his colleague, but there wasn't the luxury of time to dwell *because Pendle's people were closing in.*

An arc of flame trailed across MacGillivray's retinae: a Molotov cocktail. Hypnotised, he watched it smash against the NSSA lorry passenger door, erupting into a sheet of flame, burning tyres releasing toxic, blackened fumes into the air; the unexpected rear-guard attack throwing the NSSA crew into disarray.

*Whoever had done the damage lobbed another Molotov towards the bazooka's ammunition box.*

'Move!!!' shouted the panicked bazooka man.

*Their angel had lost the element of surprise.* Now NSSA gunfire turned in the thrower's direction.

They say fortune favours the brave, and it was indeed fortunate that the munitions chose that moment to blow. It added an addendum of flame to the lorry, causing concussive waves, forcing the invading NSSA team to retreat. *Their angel now had a chance to escape.*

MacGillivray thought he saw a boy sliding on muddy grass, disappearing into a twitten–*Ash? Run, you heroic fool. He had bought them precious moments.*

Through the incendiary noise and gunfire, MacGillivray can hear an engine fire up and can see Holroyd's familiar profile in the only police vehicle remaining.

Holroyd pulls up, and MacGillivray needs no invitation to get in.

Holroyd guns the engine, and MacGillivray looks back at Grey, who is hunched by the entrance door and providing fire to cover their getaway.

As Holroyd approaches the police car park, MacGillivray can hear Pendle above the crackling of flame, hollering orders at his team to shoot at them - as if they needed encouragement.

Bullets whine, but they're too late; *Holroyd had made it,* forging a race line arc onto the road, clipping a tree, leaving a broken red taillight littering its base.

MacGillivray's heart sang at their small win. 'Yes!' he shouted. Gaining a lopsided grin from Holroyd.

---

'WE'VE COMPANY,' said MacGillivray. It hadn't taken Pendle long to get mobile and give chase in an NSSA Land Rover.

'I noticed.' She said, glancing in her rear-view mirror. 'Where's your drop-off point?' asked Holroyd.

'Here.'

Holroyd turned the Rover to the side, applied the brakes, using a parked van to hide her brake lights.

'I'll find somewhere to hide until they pass. Thanks, and good luck.'

'Just stop them,' she said.

MacGillivray nodded, slamming the door as Holroyd sped away, almost losing a finger on the wrenched door handle.

MacGillivray crouched beside a dilapidated caravan in someone's forecourt; the chasing NSSA vehicle races up the residential road, white scything headlights threatening to expose him.

He risked a look the other way. Holroyd had slowed, damping the brakes, deliberately flaring red lights, dangling a carrot. Pendle's team sped up and hammered past, oblivious to MacGillivray.

*Holroyd's ploy was working; now he had to make it count.*

38

# Holroyd

Holroyd's Rover smashed through extravagant puddles, taking a single carriageway south out of town. Even though there would be barricades of a roadblock not far away, she felt sure she had bought time for the detective and could afford to give her self-preservation more attention.

But, up ahead, there's the bonus obstacle of a tractor dredging the road of tree debris. Her wipers work stoically to clear the mud thrown up by the heavy machinery, but she has to slow down, allowing her pursuers to gain on her.

The road straightened, inviting her to overtake the tractor. She manoeuvred out, her driver-side wheels running the gauntlet of treacherous, pooled water and fallen trees on the other lane. Her Land Rover swung from side to side, rattling against the tractor's side, its large tyre black marks a souvenir against her side's officious blue, white and yellow.

Large white blocked NSSA letters filled her rear-view mirror. 'You shouldn't crowd a lady,' she said under her breath. She registered a sign for a fast-approaching roundabout. To shake them off, she needed to gamble; she would take the third turn, hard, to trick them into continuing straight ahead.

But Holroyd's near side rear tyre clipped the kerb of the roundabout island.

'Shit.'

She fought contrary forces, trying to swing the back end around the roundabout, the vehicle's weight now her enemy. Chevrons shattered. The machine flipped and glanced off a telegraph pole. Holroyd, seat beltless, flew through the weakened, crazed front screen glass, landing in a deep ditch of water.

THE LAND ROVER LAY WOUNDED, wheel spinning, an unfortunate, innocent animal waiting for the vultures.

Holroyd's stunt had worked to a point; the pursuers had been fooled into going straight ahead at the roundabout. *But it would not be long before they returned.*

Across the road, the tractor had come to a halt, engine chugging.

She pulled her soaked blonde hair back from across her face, better able to see the Good Samaritan climb out of his cab to help.

She shouted, *'Go, get away,'* The man, looking mystified at her strange response, then faced the approaching vehicle with growing mystification and trepidation.

Pendle's Rover braked, crew exiting. Pendle blared his instruction, 'Shoot him, Matheson.'

Through rain-revived grass, she saw the tractor driver raise his hands in startled appeal. She heard him utter, 'Now, hang on a minute,' disbelief woven into his words.

Matheson's bullet struck home at chest height. The man dropped, dead weight, his arms shaking like a dog in from the rain, falling out of Holroyd's sight. She had a diamond heart, but she would carry a chip in it for his needless death, her coldly professional indifference turning to hate for the killers of this innocent man. *If she ever got out of this mess...*

Holroyd ducked and heard the splash of Pendle's team's footsteps. Then she saw their guns held hot and ready, coming into her limited view.

But they had yet to see her, hidden in the water-swollen muddy ditch.

Pendle and his people moved nearer to the police Rover, stiff with the wariness of retaliation, finding none from the empty, fallen engine.

Holroyd had lost her gun in the crash; all she can do now is to turn and try to crawl along the ditch. Her left arm is broken, complaining, leaving her right arm to do the lion's share of the work.

'Holroyd,' she turned in the ditch water to look up at Pendle defiantly, but she had blood streaming from a head wound; aided by rain, it obscured her sight from her right eye.

'It's just her, sir,' said Matheson, pointing his gun at Holroyd.

'Tell us where MacGillivray is, and we'll go easy on you.'

'Fuck you.'

Pendle surveyed her. 'She won't tell us anything. At least not soon enough. Finish the traitor, Matheson,' he ordered. The designated NSSA executioner gave a barely disguised smirk at the prospect as he stepped towards her, taking his time, drawing out the moment, enjoying it.

A thrashed engine announced its approach, and then shots rang out.

'What *now*?' Pendle asked while instinctively ducking.

One of Pendle's personnel fell, winged, to the ground, and Pendle's team made for cover.

Holroyd heaved herself out of the ditch and crawled to the fallen man, finding the treasure trove of his dropped gun nestling in grass and mud. But he isn't dead, alive enough to produce a knife that he swung for her, slashing her arm. Holroyd made it a brief skirmish by sending a bullet into the NSSA operative's heart with his gun.

She saw her saviour, Grey, now out of the car, dashing to use the tractor as cover.

Holroyd had Matheson in her gunsight. *But she wants him to know she's there, that he'll die by her hand.*

'Matheson,' she shouted, rewarded with his sideways glance. A microsecond's understanding in his eyes conveyed that he got it. *That was the gold she was looking for, that he didn't have time to retaliate, that it was all over.*

But it's a mistake. The shot-blocking NSSA Land Rover arrived between them. Through bulletproof glass, she caught the look of surprise on his face as he went from certain death to a reprieve in a nano-second. As he jumped into the vehicle, he stuck her a middle finger.

She still fired, more in hope than expectation. The recoil of her gun transmitting through her body to wounds old and new, her teeth clamped together with the pain. Her shot, a signal to Pendle that the odds were no longer weighted in his favour, made him shout for his team to go. The Land Rover barrelled down the carriageway, leaving behind more bodies, escaping the consequences of diminishing advantage.

GREY LOOKED DOWN AT HER, wavering on unsteady legs.

'I didn't expect you to turn up,' she said.

'I have to make amends for my fuck ups somehow.'

'Don't make me laugh - it bloody hurts,' said Holroyd.

'Let's hope DCI MacGillivray will make it all worthwhile.'

Grey slumped down on his haunches and fell unsteadily onto the wet grass. He held his hand to his side. Holroyd noticed the steady flow of blood pushing through his fingers.

'For fuck's sake, Grey, push something against it.'

'I have been. They got lucky and hit on me when I was back at the station.'

'You will have lost a lot of blood; you need to be at the hospital.'

'Well, I chose to help your sorry ass–so sue me.'

Grey lay back, laughed, coughed, then stopped breathing.

Holroyd crawled to his side. She held his face and watched the focus recede from his eyes.

'Grey!' she might have shouted or whispered his name but couldn't say.

## 39

# DCI MacGillivray

Holroyd had done a decent job of leading Pendle's posse away because he'd had the time to make it undiscovered to Marsh's place.

But had *she* got away? MacGillivray felt conflicted and wasn't sure whether he *wanted* her to escape. But Holroyd had played her part in giving him a chance to save the town. And, going by his experience, she wouldn't be easy prey, anyway.

MARSH'S HOUSE IS A LARGE, detached, modern new build, brashly out of keeping with old monied Victoriana architecture, as if intentionally holding the rest of the neighbourhood at arm's length.

DCI MacGillivray pushed open the black wrought gates and stepped into Marsh's domain: the family man, businessperson, master criminal - *and the killer of Yvette and Lucia.*

*But, he reminded himself, Marsh is the only solution he could see to save Hawham.*

His jaws clamped together, threatening to crack his teeth. His heart, a punch ball hammering in his chest. And what he had to ask Marsh almost made

MacGillivray forget the toll he'd taken on his body. He's hurting with the tiredness of lost sleep, of physical and emotional exertion, and can feel electric streaks of pain leap across his body. He could smell the reek of his old sweat and discern the grime of dirt and blood.

But the thought of what he had to do caused cortisol to course into his bloodstream, nature's painkiller.

MacGillivray knew there was a risk that Marsh might do away with him. *It used to be rare for a police officer to be killed. But not anymore; the value of a police officer's life has tumbled.* He braced himself for bullets that might rip into him at any moment.

But none came, and he made it to Marsh's front door.

The comforting weight of the Glock in his coat pocket, his right hand wrapped around it. *Would he use it for revenge? Or would he do the right thing?*

Then Sam Marsh opened his door. He was of slightly less than average height, looking somehow diminished, with dishevelled hair, eyes tired, weighed down by fleshy grey bags under them.

Despite the crisp, clean, late morning air, MacGillivray found it hard to breathe. The pressure in his head escalated, a telltale sign of suppressed grief, anger, fear - and from fighting the constant urge to want to kill this man.

As if reading his mind, Marsh, in an act of bravado, stepped outside his door.

As he had from court when he'd walked free, unphased and unrepentant.

*Isn't it easy to be brave when Marsh has several guns trained on him by his team? MacGillivray could see black metal poking out from windows.*

*And what is Marsh doing here? He'd escaped jail; he should be on the run. How far had society crumbled that Marsh could deny MacGillivray's authority over the law he represented?*

But then MacGillivray hadn't exactly been pristine, either. He'd contributed to the law's downfall with his planting of evidence. Within his maelstrom of hurt, he felt shame at letting the law down.

'What do you want?' Marsh asked.

MacGillivray wanted to kill him. *That's what he wanted. The heart doesn't lie; his overriding drive is stripped away, revealing his base desire. Underneath, perhaps, that's all there is: ugly revenge. But it can't be, mustn't be.*

'You're wasting your time, MacGillivray; I'm not going back to jail.' said Marsh.

'It's not my intention to take you in, at least not yet.'

'What's that got to do with me?' Marsh's belligerency is weakened by the hesitancy of his words, masking his desire to hear why he is needed and what power he might possess to barter with MacGillivray.

Then MacGillivray caught sight of Marsh's son Troy, standing in the shadow of his father. He had a look of cruel amusement, of knowledge of something terrible. He drilled his eyes into MacGillivray, looking for the unhealed, wanting to press hard into MacGillivray's raw nerves. 'How is the widower?' asked Troy with malicious glee. 'I never said it in court, for obvious reasons,' Troy punctuated his sentence with a laugh, as if sharing an amusing confidence with MacGillivray. 'It was kinda fun watching your wife's blood run on the green floor tiles of your hallway. She had fight, that one, I'll give her that. And to see your daughter's eyes as she watched her mother die. That was quite a buzz, too.'

MacGillivray felt the shifting tectonic plates of the emotions within his face as Troy gave him a hideous moment to let his words sink in.

'Stop it, Troy,' said Sam Marsh.

'It's your son that killed them, not you?' asked MacGillivray.

*This can't be true. So, Marsh had been protecting his son all this time.*

*And he'd tried to frame the wrong man.*

The implications that this revelation raised crowded in on MacGillivray. Short-circuiting his ability to think straight, a furtive ground for raw emotion to take over.

MacGillivray had harboured a compulsion to know the details of what had happened to his family. *To know if they'd suffered, as if the imagination couldn't conjure enough horror. And now, knowing just made him feel worse.*

MacGillivray whipped his gun out of his pocket, matched by the symphony of safety catches being released from the weapons of Marsh's army. Tears of passionate anger broke free. *Another few Pascals of pressure and Troy, the murdering bastard, would be gone.*

*He would surely die, too.*

*Good.*

*But what's at stake is bigger than him: the pressure of the responsibility of being a police officer.* He lowers his gun.

Marsh raised a hand, signalling his team not to shoot MacGillivray. *He would live a little longer.*

What MacGillivray didn't see coming is Marsh senior striding across the brown weaved carpet, slapping his son across the face, the assault reverberating around the hallway.

For a moment, MacGillivray sees Troy shrink into the fearful boy he'd probably been, a hint at his upbringing that had been foisted upon him.

'Go,' said Marsh, delivered with a cold exasperation.

Troy Marsh's face is a colour swatch of emotion. Pale white shock, with small cherry blossom patches of blood returning to the offended surface: hot shame turning into anger.

Troy's face cracked, eyes slit in profound humiliation, tight fists uncurling, to slam the lounge door behind him.

*That mad dog is going to come back and bite sometime soon,* thought MacGillivray.

Sam Marsh asked. 'If you're not here to arrest me, what *do* you want?'

MacGillivray's murderous blood is abating, and the world is returning to focus. Then he said. 'I need your help to save the town because you're our last chance to hold Pendle and the NSSA off. Because you've got the people and the weaponry to offer any serious resistance.'

Marsh looked at him, pondering, trying to maintain his hallmark arrogance.

'You must have dreamed of killing me, MacGillivray.'

Marsh's words triggered snatches of nightmares of trying to do that very thing. No matter what he tried to be in his waking moments, the subconscious refused to be hypocritical. Only, in his dreams, at the point of killing, his wife's face replaces Marsh's, begging him to stop. MacGillivray always woke, sweating, drained from the terrible strain of honouring his wife's wishes.

MacGillivray said, 'The town is in danger from the NSSA because we have proof that they are implementing a plan called *Project Deadhead*, under which they are unlawfully killing criminals. And you and your family could be next.'

'Not us,' said Marsh.

*Is this a toxic mix of arrogance and pride? Or something else?*

'We killed your family. Yet you come here to ask me to help you? That takes guts. You're a better man than me.'

'That is not raising the bar much.' *Not helpful*, he berated himself.

Marsh laughed bitterly and said, 'You tried to frame me, the wrong person. Perhaps you're not so much better than me.'

*He knows?*

MacGillivray was stunned by that revelation, had no answer to it. But he repeated, 'Can you help?'

'I can't do it. It's the NSSA. We can't beat them. They destroyed the Leighs. And look at what they've done to others. Now go, we're done here.'

'But the NSSA will finish the town. There'll be nothing for you.'

'When they close the police station, you'll be gone, and the town will be mine, okay?'

*Did the NSSA have that kind of power to promise towns to criminals? How far down the slippery slope of corruption had the government and their cronies fallen? How little morality did they have left?*

'Do you think you're safe, that they'll honour a deal? I didn't take you for a fool.'

'Get off my premises.'

Marsh's patience exhausted, he raised his hand and his team's guns, now pointed at MacGillivray. 'I shall report that I was threatened by an unknown armed intruder. Who would blame me for protecting my land? Who's gonna give a fuck?'

*Marsh is right. The shocking thing is, in a country of chaos and mayhem, the killing and disappearance of police officers are commonplace. And MacGillivray's removal would, conveniently, resolve an awkward problem for Pendle.*

*Marsh was never going to help, which begged the question*, 'Why did you allow me to see you?' asked MacGillivray.

'Morbid curiosity, and to see what I can get out of it. Fuck all, as it turns out. Now go before you force me to shoot you.'

The detective turned away in disgust, his strength sapped. He leaned a hand against Marsh's driveway wall, aware of the rain falling again, lashing down on him.

*He needed a moment to absorb his latest failure.*

Then he walked away, the splash of his boots falling on growing puddles for company.

'MacGillivray,' shouted Marsh.

He stopped and looked back.

'I've something to tell you,' Marsh said, looking oddly resigned.

*The momentum of the grandfather clock's pendulum behind him, served as a reminder that we're all marching to the grave.*

'About killing your family,' said Marsh, 'It was bad luck that they came back too soon.'

*He saw something in Marsh's eyes — remorse, maybe? Perhaps the death of his child, Tommy, had given him a new perspective.*

'And I was blackmailed into doing it. I want you to know.'

*Who'd blackmail Marsh to do something against him? He'd made a few enemies; what police officer didn't? But he couldn't think of anyone who'd want to kill him or his family.*

'To do what? What were you doing at my home?' MacGillivray asked, almost too quietly to be heard.

'The NSSA asked me to raid your place and to retrieve a file and smash your place up a little and give you a scare.'

*Pendle? The NSSA? Acting against me even then?*

Marsh continued, 'Pendle told me things he knew about, that he could send me down for life, not just me, but my family and crew. I thought I had covered my tracks and made myself invisible. But everything I had done was second class; they knew everything. It's all very *Big Brother*. For the first time in a while, I was scared. I still am.'

MacGillivray believed him. He could see terror behind his eyes, behind the hard man image. *He wanted to see Marsh as inhumane because any other way didn't fit the narrative he had nurtured about him.*

Marsh continued, 'They offered me a way out by getting some files from your place. "

*No one would be there. It'll be easy*, they said.'

'What files?'

'They were to do with someone called Staines.'

*The Staines files? What was so important about him that they'd kill Yvette and Lucia?*

'Only it went wrong. Your wife and daughter came back too early. They saw us and so could identify us. And for what Troy did, I'm sorry. It was wrong. If he'd worn a mask as I told him, we could have let them live. There's something wrong with him, I know. But he is my son.'

A different truth had crowbarred its way into MacGillivray's mind: That Marsh had been coerced.

---

MACGILLIVRAY WENT BACK in his memory to that pre-grey time before his family's deaths.

*He'd been working on a case, seconded by a friend and colleague, Mark Ableman, to the Swindon station. Mark wanted his help in investigating the death of Adrian Staines, a key criminal in the area. Not that many tears were being spilt outside of Staines' family on his passing. Only, Mark had wanted to give the case closer inspection. He'd said 'that something didn't feel right'.*

*He'd felt he was under mounting pressure by on high to close the Staines case and concentrate on other 'higher merited' cases. Mark had disobeyed them, had wanted to persevere with the case, and MacGillivray went along with it, because Ableman's instincts were usually on the money.*

*Staines had crashed into Coates Lake on the outskirts of town. MacGillivray had gone to the police garage to get the report. The flustered administration manager hadn't been keen on delivering bad news to him. She told him that an only copy, sent to the NSSA, had been lost in transit.*

*She said that without computers and a return to paper shifting, it had resulted in administrative carnage. To MacGillivray's mind, this was a big ask to put this all down to incompetence and crippled paper systems. No, this had the feeling of something contrived.*

*'Could I talk to the mechanic who carried out the check, please?'*

*'He left a few weeks' ago, uh, his forwarding address is unknown, the woman said, hiding behind the paper.*

*Convenient, he had thought.*

*'Could we run a recheck on the vehicle?'*

*The woman nervously pulled out a file and scanned the notes only to declare apologetically that the vehicle had been scrapped.*

*Again, convenient.*

*'Who authorised it?'*

*The Office manager shifted awkwardly in her seat. Aware of MacGillivray's growing agitation, she turned the sheet around so MacGillivray could see for himself. A coffee stain obscured the signature, obliterating all but the letters* DLE. *Pendle?*

*MacGillivray asked to use the desk phone to call Pendle's office at the NSSA. She agreed, relieved she could respond positively to something.*

*'Due to a lack of evidence, the case is closed,' the person on the phone stated.*

*'Who decided?' he asked.*

*'That information is confidential. Is there anything else I can help with?'*

*'I doubt it,' he said.*

*A day later, his wife and child were dead.*

'MACGILLIVRAY?' said Marsh. This caused him to resurface in the now, the weight of so many questions hampering his return.

*Had Pendle instructed Marsh to raid MacGillivray's place to retrieve the files and to allow the destruction of vital evidence? Had he promised Marsh the town for his troubles? And had MacGillivray been getting too close to uncovering Project Deadhead because of the Staines case?*

• • •

'I HAVE TO GO,' said MacGillivray, stunned, heading towards the outside of town and the NSSA's base camp.

His mind reverberated with these new ideas. Had Mark Ableman died at the hands of the NSSA? Had MacGillivray's family been killed because of their orders, too?

*Even without Marsh's help, he's going to try to stop Pendle. Even if it kills him.*

## 40

## Ash

Ash's nerves zinged. He could still smell petrol on his hands, a reminder of the excitement and the terror of what, he could hardly believe, he had just done. He had never seen his hands shake as they did now as he washed them in the kitchen sink.

His memories seemed to belong to someone else, of how he'd followed the noise coming from the police station, the sounds of gunfire, and explosions, where plumes of dust had risen above the roofs of the house. *He'd amazed himself he'd gone. Wasn't he supposed to be a coward?*

He remembered he'd grabbed his rucksack and shoved in his catapult and marbles–because that was all he had. But then he had a brainwave; he raced out into the garden to his dad's garden shed - *as he thought, a canister of petrol.* Ash poured its contents into two beer bottles. He shoved a cloth into their brown necks, hard enough to wedge them in. Then he snaffled some matches from his dad's worktable drawer.

He'd run out into the street. The rain had stopped, the grey sky's cover thinning. It must have taken ten minutes to get to the police station. But it felt like a fleeting blur of adrenaline.

The first thing he'd seen was the police building's roof caving in and flames crackling. He'd been mesmerised–*that's the word* - as his senses tried to take in the shock of seeing a police car on fire. Then he saw dark blue-dressed NSSA, their shoulder insignias clear, firing at people–*firing at the police!* There was a blond woman. She'd looked like *Miss Holroyd.* She'd been *shooting* at the NSSA. He struggled to marry up the image of the teacher in the classroom with the one holding a gun. She'd looked so different when she had dealt with Harry, but now she looked ferocious. And there'd been the police officer, MacGillivray, firing, too. They'd looked outnumbered. He'd had to do something. He'd swallowed, pushing down the fear that threatened to overwhelm him.

The little tags of cloth had begged to be lit with his suddenly shaking hands. He'd felt the fear and thrill as he'd taken three matches to light the first petrol bomb.

Then he'd thrown it.

The flame of the lit rag trailed a filament of smoke, hitting the NSSA lorry. A sheet of fire hungrily claimed anything it could set light to. Then incredible explosions came, lighting up the sky–*ones he caused.* He threw one more at the ammunition dump for the bazooka, and then the fireworks started.

Despite the terror, they were the most thrilling things he'd ever seen. *And that's saying something, considering what he'd been through recently.*

And then they fired at him.

Ash ran for his life. He'd slipped on the mud, falling hard onto his elbow, scrambling, running away.

---

NOW, in the stillness of his kitchen, Ash rubs at his elbow, remembering, checking it for damage. Then, someone walking by in the street catches Ash's eye. *He's sure it's the police officer, MacGillivray. He'd survived the attack.* Ash is pleased. *But there's something scary about him; he looks broken, like a runner doing a zombie walk at the end of a marathon. And the way he stares at nothing. What's wrong with him? And where's he going?*

Ash's boots fought him as he struggled to get them on, laces insolent. His coat sleeves defied his arms, his gloves refusing to offer the right holes for his fingers. But after what felt like an age, he was ready. Ash quickly, clumsily, opened his front door, surprised to find Steve on his doorstep, hovering.

It had not gone well with Steve when he had seen him earlier. Ash hated lies because he was sick of his parents always telling them to him. So, he hadn't lied to Steve, declaring that he'd told MacGillivray everything.

Steve had said before storming off, 'You did what? You're...you're a bloody grass!'

Now Steve stood before him and said, 'Eh, sorry.'

'Yeah,' said Ash, rushing past him, 'Something's up. Are you coming?'

'What? Eh, yeah.'

'Then hang on a second, I need to grab my rucksack,'

## 41

# DCI MacGillivray

MacGillivray made it to the edge of town, where signs pointed down a narrow track to the Hawham Cricket Club field. He followed it to where NSSA vehicles were parked on the field, an affront to the hallowed turf. Braziers were lit, fending off the cold of the rain. Even from a distance, MacGillivray caught the scent of burning logs and could feel snatches of heat in the wind.

*There must be over forty operatives here. How did we get to the point when a pseudo-police army sits on the boundaries of our town, locking us in, as a mortal threat to us all?*

MacGillivray could see Pendle leaning out of the open cab of a lorry, barking instructions through a megaphone. MacGillivray stepped forward beyond the entrance to the cricket ground into exposed space, where they could not fail to see him. He holds his hands up, hoping to leave no room for confusion for the four NSSA operatives who'd snapped into an offensive pose.

'I want to speak to Pendle.' MacGillivray shouts, causing heads to turn. Pendle is one of them, first looking surprised, then he walks over to the detective.

‘What can I do for you, DCI MacGillivray?’

‘You are under arrest for the unlawful killing of Tommy Marsh, Edward Bridges, Harry Butler and his father, for three members of the Leigh family, Keith Bakewell, and for the attempted murders of Ash Heath and James Leigh.’ He took a moment, then said, ‘And for the deaths of Yvette and Alice MacGillivray. I caution you that anything you say may be taken down and used in evidence in a court of law.’

‘I *am* the law around here, MacGillivray, if you hadn’t noticed. And I’m telling you, you’re finished, you bloody traitor!’ yelled Pendle.

‘Excuse me, sir,’ said a man in his mid-twenties, who wore communications headphones.

'What it is?’ snapped Pendle.

‘I have been listening to a local radio broadcast.’

‘And?’

The comms operative faltered, but then his courage propelled him forward.

‘They say the NSSA has helped instigate the Nation's systemic culling of the criminal elements for political gain under Project Deadhead. And that the project is in danger of being revealed for its true purpose, and that we are here to deal with those that have proof, by any means possible. Is that true, sir?’ The comms man, beyond nerves, is now looking angry.

‘It’s becoming one of those days,’ said Pendle tersely, unable to hide his tedium. ‘There is more at stake than you know. Don’t let cheap propaganda stop you from doing your job, or I swear you’ll spend your remaining years behind bars. That’s if you’re not shot for mutiny.’

The comms operator’s fear is making a comeback,

but he declares with stoic resolve, 'I am placing you under arrest, sir,' pointing a gun at Pendle.

Pendle laughed coldly and said, 'Everyone wants to arrest me; I feel flattered by all this attention.'

Pendle lifted his hand slightly. A signal one of his staff routinely acted upon; red blotches of blood appeared on the comm's man's chest from his gunfire. After a slight exhalation of air, the body spasmed.

Pendle turned before the man had hit the ground, saying, 'Traitors. We're surrounded by bloody traitors.'

Shock at the cold-bloodedness of the killing slowed MacGillivray's reactions. He belatedly pulled his gun from his coat pocket.

'I wouldn't do that, sir,' said the man, with a name badge called Matheson, now holding his smoking gun at MacGillivray. 'Drop your weapon.' The two other NSSA operatives snapped the safety switches off their rifles.

MacGillivray thought even Pendle looked a little shaken even though he managed to procure some words through his shock, 'I congratulate you on getting your message out, MacGillivray. Now, because of you, hundreds will die. Any last words before you're shot for failing your Nation?'

'What message? I don't know what you are talking about.'

'Don't give me that,' said Pendle, the corners of his mouth downturned in contempt. 'Anyway, it doesn't matter. No information about *Project Deadhead* is going to be leaving this place. Make it slow, Matheson; he deserves it for the trouble he has caused.'

With more enthusiasm than duty, Matheson

moved in on MacGillivray, signalling to two colleagues to join him.

One of the NSSA swung a rifle butt at MacGillivray. The detective's upheld obstacle of an arm cracked with the impact. Another brave soul rammed a rifle butt into his stomach, causing him to roll sideways, retching. A heavy black boot ploughed into the side of MacGillivray's body, causing him to lurch back.

'Steady lads, pace it,' said the man called Matheson.

Pain wracked his abused body; then another rifle butt gratuitously smashed into his cheek. The detective could feel blood rolling down his face, failing to thicken in a drizzle of rain. He almost welcomed the punishment for failing again, steeling himself for more blows to follow.

But none came.

The NSSA operative who had swung a boot into him shouted in pain, 'What the fuck?' clutching his face. The one called Matheson yelped, holding the back of his head. MacGillivray noticed an orange-eyed marble laying in the damp grass.

*What's happening? Who's attacking? With... marbles?*

'Leave him alone,' a youthful voice shouted across the field, losing its power over the distance and rain. The boy fired the catapult again; this time Matheson caught the flying round smooth glass with his hand.

A gun went off, a warning shot. A shout from an NSSA operative demands that the boy drop the weapon. 'Or the next shot will not miss.'

'Do it,' shouted MacGillivray.

MacGillivray saw two boys descended upon, now held by NSSA personnel.

He fumbled for recollection and found it. *It's Ash again*. Wild-eyed and shouting, 'Leave him alone!' as he and his companion are marched away.

Despite his pain and failure, he still felt warm pride in those boys and fear of what might yet happen to them.

'Is this all this town can offer? A deadbeat detective and a couple of kids with a *catapult*?' said Pendle, eliciting mirth from his team.

'You still sound worried to me,' said MacGillivray, spitting blood from a cracked lip.

The top of Matheson's uniform is blood splattered from Ash's marble strike. He turns and swings his boot at MacGillivray's head, saying. 'Shut the fuck up.'

MacGillivray could feel his consciousness disintegrating, but he hung on grimly.

*He'd thought his badge would be enough to protect him, but we're way beyond the normal rules here. What an idiot. I'm going to die.*

One of MacGillivray's transgressors stamped heavily on his cracked arm. More bone easily gave way to three hundred pounds of dense muscle. He gasped, almost blacking out with the torture, left praying for the white pain to disperse.

His brain, looking for a way out from his torment, mixed memories with reality. Braziers burned, emitting loud crackling and bangs as wood knots exploded. His mind taking him back to childhood memories of bonfire nights.

HE REMEMBERED A CATHERINE WHEEL, *pinned to a wall, swirled, sparks flying.*

*Closed off to traffic, people freely snake their proces-*

*sion through the town, dressed as Victorian soldiers, sailors, and pirates. Their tar barrels carrying fire. Flaming torches held aloft, casting silhouettes of shadow against old buildings in the black of the crisp November evening. Their bonfires, where effigies of the most unpopular character of the day would burn. He remembered yellow embers rising and dancing, the aroma of rebellious celebration heady—a frightening, exciting, medieval sight.*

HE CAME BACK to the now; The NSSA's blows changing him into a bruised mass of pain, his receptors packing in, his consciousness on the verge of giving in.

*Perhaps he's the effigy being destroyed.*

*But what's that sound?*

IN SOME DARK logical recess of his mind, MacGillivray comprehends that he's confused from jumping between the now and the past – *so what does he trust?*

*But there it is again.*

*Drums.*

*Now there's chanting.*

The brutality dealt out to him had stopped. The merchants of his pain were all looking in the same direction.

MacGillivray tried to sit, leaning on his broken arm and falling back down again. He rolled painfully to his good arm's side, to lean, to follow their gaze.

An aura of people with handheld flaming torches projecting a phosphorescence that emanated in the late afternoon's gloom. Tractors came into view, their lights spread beams, creating a melange of shapes and shadows in the dark underbelly of clouds. The tractors are pulling trailers, and people are hanging off handholds.

*It's like an army of ghosts metamorphosing out of the half-light.*

*What other-worldly image had his punished imagination conjured up now?*

They were pooling at the roadblocks, spilling out, spreading across the boundary edge of the field. The drumbeat, distinct, beaten to a medieval tune–one-and-one-and-one-two-three – that had an echoing defiance in every pulse.

An unquiet rhythm to rebellion.

PENDLE LOOKED ENERGISED, shouting to his people. 'Hasn't it been easy to roll people over in this country? It'll be fun to have some resistance for a change.' Pendle's people machine roared its approval.

*For Pendle to talk like this means he must've truly lost it.* MacGillivray thought. *How can you reason with madness?*

However, Pendle looked alarmed when twenty or more of the town's contingent started marching into the no-man's-land, where MacGillivray and Pendle's team were. Pendle evidently viewed it as an opportune moment for him to back away to the relative safety of the NSSA lorries and the protection of the majority of his workforce.

MacGillivray's assailants now had their weaponry trained on the newcomers. MacGillivray thought those people coming had an aura of defiance about them. *The guns helped with the impression–of these unconventional, avenging angels.*

'Step away from him, or we shoot.' Those words give hope to MacGillivray - but a sickeningly familiar voice speaks them, too.

*Sam Marsh appeared to be his saviour. Marsh had changed his mind about helping.*

MacGillivray grimaced at his inner conflict. His life had become all about the sliding down of snakes. And here's one sweet, evil ladder of hope.

MacGillivray looked at Matheson and his team, at how they bandied together, always covering each other against any threat to their tribe. Their nostrils flared at the whiff of blood in their noses, their senses intoxicated with the brutal control over a human being. Him. And they seemed to have a taste for more. They clasped rifles, aching to finish MacGillivray and his newly arriving companions, waiting for Pendle to give the go-ahead.

Pendle's voice boomed through the megaphone: 'Retreat.'

The NSSA operatives walked backwards, their rifles held, covering each other. Matheson is looking disappointed, yet still swings a kick at MacGillivray's broken arm, satisfied with the agonised yelp of pain he gains.

MARSH TRIES TO HELP HIM, but MacGillivray shrugs him off, stumbling back to the line where the sound of drums came from.

'What changed your mind?' MacGillivray asked Marsh. 'Did the title of *hero* appeal, after all?'

Marsh's smile had the lean shape of bitterness. 'I'm no hero, MacGillivray. And you're right; there's no deal for me, no town to control. If the NSSA wins, there'll be nothing left for anyone. I'm here to make a stand for my family's survival.'

MacGillivray nodded, understanding that much. He asked. 'Where have all these people come from?'

'We got word out.'

'You did? How did you contact so many people so soon?'

'That'll be Sid with his ham radio–it's been useful during these post-internet days.'

MacGillivray still didn't get it: *ham radio - isn't that ancient tech?*

More out-of-step marching folk were arriving, following the wheel tracks of arriving transport.

MacGillivray's eyes adjusted to the gloom, recovering a little from the assault he had endured. He could now make out handguns, shotguns, rifles, light refracting off petrol bombs, heavy metal shields, chains, bricks, and cricket bats being waved in the air. *And is that a crossbow, for heaven's sake?*

'They came out for you, DCI MacGillivray. It's you who is the hero, you mad fucker.'

MacGillivray turned. His broken arm pulsing, but not entirely lost in the blanket of pain that covered his body, his face a red ruin, but he lifted his good arm to the crowd.

They cheered. Gunmetal, lovingly oiled, shone eagerly for retribution. MacGillivray saw the angry young that had found something to strike out against. The old army, fossil fuel war stalwarts, wetting their lips as muscle memory returned, with a sense of worth and camaraderie unpacked after years of hibernation.

MacGillivray looked around at all these men, women and, bloody hell, children. Brave, ready to deliver a reply to a government that had betrayed them.

They'd wondered if their town had been cursed, only to discover something more tangible had been damaging them, that the town had been ruthlessly picked for culling by this organisation *because we had lifted the lid on Project Deadhead.*

MacGillivray turned to face the NSSA in their dark regalia. He thought a British uniform symbolised courage, honour, and respect. *Yet these government operatives were a threat to our town and our freedoms. Something had been warped here and dressed in the darkest blue.*

'What do they say about good men?' He asked under his breath.

A man he didn't know heard his question, put his hand on MacGillivray's shoulder and said, 'The only thing necessary for the triumph of evil is for good men to do nothing. That'll not be you today, officer.'

'That'll not be many of us today,' MacGillivray replied, noting the increased swelling of the town's ranks.

MacGillivray stepped forward ahead of the massed hordes.

The drums stopped. The palpitating tension of the silence beat a louder tune, demanding an answer from the NSSA to its challenge.

Lights from vehicles catch thin rain in their bright channels in the late autumn day as it draws up its night blanket on the land, as the sun goes to bestow its gifts elsewhere on the planet.

MacGillivray turned to cross no-man's-land. To where he had left blood, teeth, and dignity.

'Where are you going? Are you mad?' asked Sam Marsh.

'I've got to speak to Pendle, or people will die.'

MacGillivray's words spread like wildfire through

the rebels. A reverential quiet descended as he made his way out again, to the baron strip of land in the middle of the cricket field. He wants to make an explicit statement of no fear and invite Pendle to meet him. *But will he come?*

PENDLE HELD his megaphone in one hand and a handgun in the other. It felt like a long time while he seemed to weigh up the situation; he must know that he will look weak if he does not come out, so he walks towards MacGillivray, stopping short by a few metres.

'Our government could do with more people like you,' said Pendle, grudgingly. 'People that can lead. But you're dangerous, too; you're deluding the town into believing they can win. Can't you see you're leading them to their deaths?'

MacGillivray spoke, dried streamlets of blood cracking at the sides of his mouth, red flakes being carried away by rain. 'The government thinks it can ride roughshod over the people, to treat them as acceptable collateral losses, in this economic war favouring the fortunate, the ones borne into wealth and easy opportunity. It's not right that the government executes its citizens in the name of *Project Deadhead*. It's just a disguise for power gain, a betrayal of the people it should protect.' MacGillivray took a breath. 'Who decides who is guilty? Who's worth less to society? Who should be culled? Hawham will be a symbol of a stand made for all. We refuse to be victims of persecution. Project Deadhead must not continue.'

'Excellent speech, MacGillivray. As I say, I think you're wasted. But what you think is an illusion because those with real power rule, like it or not, *that* is the reality.'

'I am sorry, Pendle, it's over.'

Pendle raised his megaphone and spoke to the Hawham hordes. 'You are breaking a government-sanctioned curfew — and you are breaking the law. Turn away now, return to your homes, and the Government will look leniently upon you.' A jaggedly metallic megaphone punctuated his sentence, fuzzing the edges of his words.

Pendle's lying words met a homogenous block of subversion that manifested in a defiant roar from the rebels, with the banging on vehicles, drums, and makeshift shields – growing into a crescendo.

*Bosch would have had a field day painting this scene,* thought MacGillivray.

MacGillivray did not need to be an expert in reading body language: this uprising sends an enervation of unease through Pendle.

But Pendle lowered the megaphone, leaned forward and said, loud enough for only MacGillivray to hear, 'I am going to tell you a fundamental truth about me, MacGillivray. Deep down, there's a place where I hold a dark wish. For the masses *not* to stop, I want them to *defy* me. I want to punish them for wasting my time with their petty grievances and sub-optimal ambitions. I want to do them all a favour and end their cheap existence. So, fuck the lowlife wasters of this Nation. Let's get on with this.'

Pendle raised his left arm.

THE RESPONSE to Pendle's order is the percussion of a bazooka ejaculating its shell.

Pendle looked across at MacGillivray and laughed at him, mocking MacGillivray's shock.

'It's a show of faith, trusting the shell's trajectory in this opaque weather. We aim for soft targets, though we can never be sure what we'll hit. This is insane, MacGillivray; we can agree on that, at least. But it's you that can stop it.' His eyes carried behind them a blighted light. *He's utterly deranged and thrilling in the sheer power.* Pendle laughed at the synchronised looks of disbelief on the opposing assembled hordes as their heads turned to follow the smoke trail's path.

Rooks vacated the trees in the distance, cawing their complaints for this rude disruption. The shell brutalised the land and synchronised with groans and screams from the crowd.

'It hit the park, MacGillivray, a zero-collateral target designed to deliver measured shock value to the rabble. It looks like it's working. Are you giving in?'

'You're going to cull us, anyway. What do we have to lose? They know that.'

'True,' Pendle said, then ordered another shell to be fired. It whistled through the air. The designated target is a church. A rumble unsettled the earth after the successful strike before the sound reached out from the tortured edifice. The resulting fire burned enough to illuminate the surviving parts of the structure in stark Gothic relief.

Pendle said, 'Did you know, MacGillivray, that climate change with its tragedy has caused people to face their mortality, so many have turned back to religion. So attacking a symbol of spirituality can help to break some people's resistance; therefore, shelling the holy building will be psychologically damaging to many.'

Pendle held up his hand, an order to cease firing.

MacGillivray looked about him. Artillery smoke did an improvisational jazz dance with the wind, adding a strange elan to the scene of aggressive hot metal—a contrary beauty against the calculated terrorism.

An invoked fear rippled through the massing ranks as they shuffled back a step. *A small amount of land gained, but it is telling.* MacGillivray may have won the first psychological round, but MacGillivray hated to admit that Pendle could have won the second, taking away their momentum.

Pendle climbed onto a Land Rover. No doubt, calculating the imagery would be powerful and intimidating.

'You're staying silent, MacGillivray. So, you're not surrendering? Well, fuck it, let's hit the railway station.'

Another of the bazooka shells on steroids left the metal tube of the shoulder-balanced weapon. It hit the long-standing historical building of the railway station, causing part of its façade to disappear into brick rubble and dust particles, leaving the Victorian structure seemingly aggrieved at its damage.

The following shell warped railway lines. A veil of dust, damped down by the light rain, allowed a glimpse of flames that stretched for the sky from ruptured gas pipes. A damaged fire hydrant gave off an effervescent thrust of water.

MacGillivray would not have blamed the townspeople if their resolve had melted. *But they weren't running.* He scanned, gauging their faces. *Yes, they gasped in shock, but there is a wilful solidarity, and they're still united against Pendle's attack on their town.*

'Last chance to save a few lives,' said Pendle, as if addressing an errant child before administering a punishment. Pendle looked thoughtful, then said. 'A few? It's strange how the more lives we take, the less meaning they have.'

*It seemed to MacGillivray that Pendle had gone far along his nihilistic path. Yet he sensed Pendle's underlying contradiction of his nervousness against his power lust. He's procrastinating, hesitating to commit mass murder. Pendle had ordered killing before–so why is this different for him now? Because he had never faced it, that's why. Because he's in the front line, and he could die.*

*But the stakes had to be high for Pendle to risk being here; it must be 'do or die' for him. Because if he failed, the people above him would not be happy.* MacGillivray could imagine them demanding to know how Pendle had played this so badly, to give so many people cause for unity, to let information slip out about *Project Deadhead*. And to risk bringing down the culpable party–the government.

*Pendle would be finished.*

*That's why Pendle would, despite his faltering, go through with mass murder.*

Then MacGillivray heard Matheson, Pendle's pet psychopath, shout to Pendle. 'Ready when you are, sir.'

MacGillivray could swear he saw Pendle allow the last vestiges of what had constrained him to fall away, see it in his face, as he closed his eyes: breathed in and then slowly breathed out, savouring it.

Pendle shouted viciously at MacGillivray. 'I bloody warned you. To hell with the consequences,

MacGillivray. I can get away with anything - and isn't that a good enough definition of true power?'

Pendle's team raised their automatics, and the person with the bazooka tilted the weaponry towards the crowd.

MACGILLIVRAY LOOKED behind him along the line. He saw people who'd had enough and now prepared to put everything on the line. The terrible courage in people who knew they may have arrived at their last moments on this earth.

People spoke of the cult of the individual. Of the rise of the selfish.

*Not today.*

*Not in this place.*

He lifted his arm, ready to set these people against Pendle's force, dreading the losses to come.

*But if he survived, how would he live with all this blood on his hands? And the guilt he felt for his family?*

*Yet, again, he realised he had missed the point. People could determine for themselves what they are prepared to give up. To stand for what they believe in. For what they love.*

*That's what his wife had done.*

*So how dare he be so arrogant to presume to control their choices? MacGillivray felt constrictions in his body ease, as he felt reconciliation with his past, at last.* And he gave a wry smile at the timing.

He prepared himself to give the order.

---

MACGILLIVRAY FELT THE AIR MOVE, of a premonitory sense of pressure.

Then the grey sky filled with the sound of a helicopter, its blades' sword dance indiscriminately slashing through the rain, the two factions as humans press-ganged into protecting faces with their arms against redirected spears of water.

Groans of alarm rose as one in the crowd.

He could hear a few cheers from the NSSA team, assuming the air support must be for them. *Are they right?*

A side hatch opened on the helicopter. A person with a megaphone addressed the assembly.

MacGillivray said with surprise, 'Is that Peterson... the deputy PM?' to no one in particular.

As the 'copter came closer, he saw he was right—it's the minister. With her gun-toting support from the hatch of the aircraft.

*They're going to finish us.*

## 42

# DCI MacGillivray

Peterson's next words stunned MacGillivray, and by the look of him, Pendle, too.

'Pendle, this is Minister Peterson; I command you to *stand down.*'

Pendle stared in wide-eyed disbelief. He brought his megaphone, in small increments to his mouth and shouted, 'Stand down? Stand bloody down? Why would I stop when I've got the solution in the palm of my hand? Our hands!' His words were almost lost in the electric low noise of the hovering machine.

Peterson repeats the order, 'Pendle, stand down, or we shoot.'

A person behind Peterson accompanied her instruction with a rifle shot from the helicopter hatch. A bullet hit short of Pendle's feet. Sounds of shock and confusion emitted from both sides of the field's divide.

For a long moment, Pendle appeared to wrestle, his face now contorted with his frustration.

*Is he considering attacking us, anyway?* Thought MacGillivray.

• • •

'STAND DOWN,' said Pendle, resigned. Then he threw the megaphone to the ground.

The crowd moved back as the helicopter landed. They watched in astonishment as Peterson, someone as close as many of them would get to celebrity, stepped out of the craft.

Like a pack of dogs closing in for the kill, Peterson and her armed escort marched to Pendle.

'You played me. Am I just a sacrificial piece on the chessboard of your game?' asked Pendle of Peterson.

'I had contingency plans, if that's what you mean,' she replied. The tone of indifference in those words startled Pendle.

'You are under arrest,' she said.

Pendle looked like he'd been unplugged, as if he'd hardly heard, as the NSSA placed handcuffs on his wrists, their metal tight and cold.

*Petersen carried out these actions before the watching crowd, ensuring word would spread that Pendle alone would be responsible for this aberration in Hawham and not the government. And not Peterson.*

*Pendle looked humiliated, cold, and tired. Unthinkably defeated. A frown framing his eyes in the darkness of his betrayal.*

It all fell into place for MacGillivray.

*Petersen had ordered Pendle to seal off Hawham, then soft target shell it, enough to put fear into town, to show the people the threat's real. Then, when she intervened, the people of Hawham would be grateful. The government would have someone to pin the blame on: Pendle.*

*Pendle would be the villain.*

*She would be the hero.*

*She's brilliant. And dangerous.*

## 43

# DCI MacGillivray

MacGillivray and the people of the town had glanced over the precipice and seen death staring back, yet they had stood up to the NSSA forces. And, by a strange turn of events, had survived.

*Would Project Deadhead continue?* He guessed they would suspend it until this episode with Pendle faded. And perhaps all he had done was slow down this behemoth that Peterson was driving.

He took a deep breath, smelling the acrid residue of adrenaline waste from his lungs. Then people came out of shock, hugging him, slapping his back - inadvertently jolting his broken arm. He peered into the clearing and looked towards the whirling of large blades speeding up, lights flashing. Pendle cut a forlorn figure being led away, stowed on board like luggage.

Four armed NSSA personnel approached MacGillivray. One raised an eyebrow at his state of injury and asked him if he's fit enough to follow them to see Minister Peterson. 'I can manage.'

MacGillivray could feel the crowd agitate around him, wanting to guard him. 'It's okay,' he said, using the placating action of his hands, hoping his confidence in his safety was well placed.

UP CLOSE, the Minister for Crime, Policing and the NSSA seemed slighter than he expected, though her eyes looked hard and brittle as a deep winter.

'DCI MacGillivray, the government and the country owe a great debt of gratitude to you for stalling Pendle until we could apprehend him. We are all grateful for your help and courage in bringing him to justice.'

'It's my job,' MacGillivray said.

'It won't be forgotten.'

'Not by me,' he said.

Peterson managed a tight smile, her skin fighting against this unnatural position. *Perhaps she'd picked up on the ambiguity of his words.*

'You're brave or foolhardy, and assessing your recent performance, I am going with the former. I'm always looking for the right people with integrity and intelligence to join my team.'

MacGillivray couldn't keep the incredulity out of his face–*is she trying to buy him off?*

'I'll consider it.'

'Good. Is there anything we can do for you in the meantime?' Peterson asked.

'Would you consider releasing Tim Salisbury, Ash Heath and Steven Taylor?'

'Ah, the broadcaster and the brave boys. Yes, I can do that as an act of goodwill.'

'Safe journey back,' he said, trying to avoid a hint of sarcasm.

Peterson couldn't hide the irked tone of her voice.

'And don't *consider* my offer for too long. Goodbye, DCI MacGillivray.' *There it is: the threat.*

Accompanied by four guards, she hastened to the chopper, rifles covering her back. *Trust in the people isn't at a premium, then.* Thought MacGillivray.

'PETERSON.' said Sam Marsh, pushing through the assembled throng.

NSSA personnel snapped to the defence of the deputy PM, weapons ready.

*What's he going to do?* Thought MacGillivray.

Peterson stopped and took in the speaker; impatience signalled with a roll of the eyes.

'Yes?' she said.

'You had my son killed.'

'I am truly sorry for your loss, but, no,' she said with barely disguised contempt at the intrusion, 'Pendle has been acting as a rogue element and carried out the killing of your son along with other actions without our knowledge for his agenda. But rest assured, the justice system will deal with him. And, any deals Pendle brokered are to be reviewed,' she said pointedly.

*Is Marsh to be bought off, too, or is he exposed and now a target?* Thought MacGillivray.

Marsh also looked like he had that impression as he raised his gun. Troy, along with the Marsh crew, copied him. The snap of firearms by a phalanx of NSSA operatives strained the atmosphere, guns tensely trained on Marsh and his cohort. MacGillivray noticed that the circle of attending town's folk was now widely retreating fast.

'Are you ready to die now?' she asked.

'Are you?' replied Marsh.

MacGillivray's first instinct as a police officer is to stand between Marsh and Peterson; he does so, saying to Marsh, 'It'll do no good for you to die.'

'But she'll be dead.'

MacGillivray can feel the cellar door opening within him, and a dark thought makes itself known: *He could step aside and allow Marsh to carry out his act of self-immolation. And Peterson would die. Perhaps even Troy Marsh might get what is coming to him. So many acts of revenge could be sated.*

*But he could not live with it. And it would be risky to have so many innocents caught in the crossfire.* He kicked the idea back down the dank steps of his heart and slammed the door shut.

Marsh lowered his gun, and his team slowly followed his example.

Except for Troy, who strained at invisible leashes.

'Troy, don't,' said his father, more broken than authoritative.

Troy succumbed, high-standing veins barely subsiding.

'We can live to fight another day,' said Sam Marsh

'Will you stop talking bollocks?' Troy snapped.

Peterson said, 'Once again, thank you, MacGillivray. We'll be in touch,' as if it had been a minor interruption, then she resumed her exit.

MacGillivray is impressed, despite himself, with her coolness in the face of death.

He watched the 'copter take off, disappearing into the darkening sky.

Within moments, all that's left are scars in the green field, of tyre tracks, scorch marks, and the trails of people eager to celebrate still being alive.

## 44

# DCI MacGillivray

MacGillivray stood in Sam Marsh's front room.

The detective's right arm is in a cast, and he had a few cracked ribs from being shot; his bullet vest had saved him from their impact. He also had a few broken teeth that were causing him discomfort.

That antique grandfather clock still dutifully ticked off the minutes in Marsh's hallway. But MacGillivray felt that the quality of time had changed for the better somehow, but he needed to reflect on why that is the case.

MacGillivray said. 'I'm grateful for what you did for the town, Marsh, but your amnesty is over.'

'I get that, and let's say I am a changed man,' said Marsh.

'I hope so.' *Do people like Marsh change? How often does anyone really change?*

MacGillivray looked across at his wife, Sadie Marsh. *She looked unreadable.* And as for Troy, he doubted he could describe him as a *changed man. Nearer to tempered bloody steel. He's the heir to the throne, or maybe an ogre in waiting - and it's Troy who has the blood of MacGillivray's wife and child on his hands.*

Troy looked at him, tried to smile his cocky smile, but a hint of doubt sat in his eyes. Perhaps he knew MacGillivray would catch up with him soon.

---

MACGILLIVRAY STOOD TENTATIVELY next to Crozier's bed in the cramped hospital room; she lay there, bandages around her head and shoulder.

'Gezellig, right?' she said.

'Yes, cosy.'

'Ten days in a coma and a bullet wound, you had us worried, DI Crozier.'

'I'm back so don't worry; the doc said it all looks positive. Now, what have you got there?'

MacGillivray had a small pile of books within the crook of his unbroken arm, placing them on Crozier's hospital bedside table.

'I am honoured,' said Crozier, aware of MacGillivray's love of his books.

'Don't damage them, Or I will, of course, hunt you down.'

She laughed. 'Look at you. Yellow and purple are not your colours, MacGillivray, and what will happen to your teeth?'

'Titanium implants. It'll take a few months,' he said, holding a hand up to his face, opting not to touch.

'You're looking well, considering.'

'I was lucky. If the explosion hadn't blown me into the back office, the roof timbers would have crushed me.' She gave her head half a shake, half a shudder. 'What about Tim?'

'Peterson kept her word and released him without charge. He was unreasonably apologetic to me.

Thankfully, Sally was doing a good job of telling him so.'

'And Hanbury?'

'He has a minor concussion but will be fine.'

'Thank goodness. And what about Ash?'

'He, too, has been released without caution. And he's alright, considering what he's been through.'

'Good. I hope he has not developed a taste for pyromania.'

'Well, I think he is seeing an old flame called Angelica.'

Crozier half laughed; half groaned at MacGillivray's pun.

'Ash has told me he wants to be a police officer. I think he'll make a good one,' said MacGillivray.

'You mean if robots haven't taken us over by the time he qualifies?'

'Well, that pessimism is not what I expect of a future DCI. But on that topic of robotics, the Peel Bot trial has been suspended and is under review.'

'For now, but they are coming, aren't they?'

'Yes, almost certainly,'

MacGillivray paused momentarily, then asked, dropping his voice, 'Did you remove the evidence on Marsh?'

'Yes,' said Crozier, tensing. 'I knew you had planted it and weren't in your right mind. I knew you would regret it later and that it would do a good job of eating you up.'

'You're right. I regret bitterly what I did.' Then he said. 'But why did you take the risk? You could be charged with tampering with evidence.'

She laughed. 'Forget it; it's our dark secret, and we've now got our own mutually assured destruction,' she said, trying to dissolve the awkward moment.

He laughed, the sound is still alien to him.

'You know, Sam Marsh must have thought the NSSA had removed the evidence to save him. He was wrong; they were not his guardian angels, and he was always targeted to be destroyed.'

'Do you think they will still try to wipe him out?' asked Crozier.

'Too much has happened; Peterson could not afford the risk right now.'

'Especially with Pendle taking the hit for it all.'

'It looks that way. However, it'll be a cushy number in an open prison, but he won't be coming out anytime soon, I think.

Do you know that Grey was not so lucky?' asked MacGillivray.

'Yes, I heard something from the nurses; it's tragic, considering he was on a path away from this lifestyle. And what about Holroyd?' asked Crozier.

'I don't know, she's the NSSA's problem now, but word has it she has disappeared.'

'You don't sound happy about that.'

'How can I feel positive about someone like her after what she did?'

'She helped save us, MacGillivray. That must be worth something?'

'Maybe.'

'And Peterson gets away with it?' asked Crozier.

'For the moment.'

She took one of her trademark deep breaths. *What's coming now?*

'It's okay to talk to someone, you know, about what Lauren Bakewell mentioned, about what has happened to you.'

'I've had a preliminary meeting with a psychologist — Mrs Bakewell, in fact. She is thinking of starting a private practice.'

'That was quick work by you and good for her. Anything you can share?'

'Everyone that I had ever tried to protect was dead, and that I need to guard others, to give to others that which I can't get for myself. That seems to be the gist. And I think I am ready to face it now.'

'You're an emotional junkie and a wreck, then.'

*Only Crozier could say that and make it sound like care and funny at the same time.*

MacGillivray kneaded at the red skin between his thumb and forefinger. But not as unkindly.

## 45

# Holroyd

The winter of heavy rain and floods gave way to soaring temperatures that crisped greenery, threatening to combust trees, making pavements hot enough to singe flesh.

Shades of relief came from the south-westerly breeze that blew off the sea, making the outdoor heat just about bearable in the early hours of the day. *Much better than the heated claustrophobia of the inner towns and cities. And it's a beautiful part of Cornwall,* thought Holroyd. She wore black wraparound sunglasses, a long-sleeved red tee shirt, and light black trousers. Holroyd liked the heat of the sun, but only sparingly.

*East Looe is one of two pretty towns, the other being its west side sister, but it's living on borrowed time. The trouble is the rivers and seas are rising. Nature's clawing back the land from the stewards that had defecated on it so profusely. Give it a little longer; this place will be as good as uninhabitable, submerged under so much water.*

Holroyd made her way along the harbour to a pub, The Call of the Sea, enduring its ironic countdown to watery abandonment.

*Staying in hiding must've given Jacobs a hefty dose of cabin fever because he had finally given in and started frequenting a local bar. Isn't people's company supposed to be good for your health? In Jacobs' case, it would prove false, a deadly mistake. Or maybe it's his convoluted way of facing the inevitable.*

JACOBS SAT ALONE in a dark corner of the bar, slightly swaying, several empty pint glasses in front of him. Despite his drunken state, he seemed to recognise death calling when he saw it, going by the way he reached for his glass. He pours beer into his mouth like it is courage, a little spilling on each side of his chin.

She slid into a seat alongside him.

He sighed, resigned, and asked, 'What'll you have?'

Holroyd said nothing.

Jacobs caught the eye of the notably unsurprised bartender and signed for two beers. *It's easy to catch the bartender's attention with no one else in the bar this early in the morning.* She thought.

'How did you find me?' Jacobs hunched over his beer as if trying to decipher one of life's mysteries.

She ignored his question.

His sentences run into each other, lacking the politeness of a gap for a reply. *He isn't rude, necessarily, just drunk and scared.*

A thought seemed to occur to him. 'Aren't *you* on the run? Why are you chasing me? No... wait..., Grey died, didn't he? I'm sorry. I didn't mean for anyone to get hurt. So, you're motivated by vengeance.' *He seemed darkly satisfied with his calculation. Maybe finding a reason for his demise would make some sense, as if there was comfort in understanding something about his existence.*

Jacobs tipped the glass to his lips and drank while tears broke. *Is it possible to sob and drink? That's a neat trick.*

'My marriage has been a car crash. She's been carrying on with at least one neighbour. I can't blame her. Hell, I'd cheat if I was in her shoes. And I'd cheated on the job, taking bribes.' *The ubiquitous confessional. Especially for a bastard who had worked both for a criminal and fed the NSSA info.*

Holroyd pulled the gun, with a silencer, from her shoulder holster. She laid it on her lap, pointing it towards Jacobs. *She knew she looked bored, but she couldn't help it. The man's rambling, getting on her nerves. But aren't we entitled to a few last words?* She thought she could afford them as *he might just say something useful when death is seconds away.*

'I wanted to be finished with all the dirty deeds. Only it never is, is it? You can never get away once you start. They have you.' He wiped a hand across his face.

The bartender slapped two pints of Tribute beer on the table, leaving splash marks on the placemats and table. Jacobs passed some money to him, rewarded by a small smile at such a large tip.

'I had felt the panic, the station under threat of closing, with the news of robots taking over our jobs. If I'd lost my job, I would've been finished. I needed some insurance, a buffer. I couldn't think straight because of it.'

Jacobs' speech, quickening, thickening.

'It started when I borrowed money from a loan shark, not realising it was Leigh's money. Later, when the amount had escalated. They gave me a way of paying it back. All I had to do was turn a blind eye to one of their jobs. It seemed harmless enough, as they were only raiding a bookie. Can't those blood suckers afford it? And those fuckers had helped get me into my mess in the first place.' His eyes briefly fired with self-righteous, self-pitying anger, and then they dwindled. He bunched his shoulders up, gathering himself to continue with this protracted, self-administered epitaph.

'Then they pressured me into filtering information to warn them of police raids. Because if I didn't, they would tip off my boss that I'm *bent.*'

Jacob's face froze at the use of the word. It was as if saying it out loud released the poison of realisation of what he had become. *Own it,* she thought.

'I thought I could squirrel enough money away for the dark days to come. Hadn't I been right?' another bitter laugh. He drew heavily on his beer. Holroyd thought she detected a smell of acrid urine.

*Enough. This is cruel.*

Jacobs looked up at the barkeeper, some light of comprehension filling them. 'He doesn't look surprised. He told you I was here, didn't he? But I'm too confused to work out how you found the right town. But it doesn't matter, right?'

Holroyd didn't feel inclined to let light in upon magic.

His breaths were shorter, shallow.

'Oh...wait... my wife told you I'd always wanted to live here. Of course she did. I thought she took no notice of what I said.' he gave a bitter, reedy laugh, trailing off into a stifled choke.

'Pendle contacted me, you know. He told me he knew everything about what I'd been doing. He knew *everything*. He said I'd be clear if I did this one thing. To provide a list of key candidates to be placed on Hawham's kill list. Imagine having the power to write someone's name, and they would be *gone*. I stuck a few people on there that shouldn't be, just to get my own back on them.'

Holroyd wished she hadn't heard that. The thought that she might have disposed of someone innocent registered a reasonable-sized seismic blip on her conscience.

He took another generous gulp of his beer, realising that that revelation had hastened his doom.

'I'm sorry. So sorry...for everything...will you let me live? Say something, for pity's sake.'

Holroyd favoured him with a few words. 'Believing you are sorry, and forgiving don't always live in the same neighbourhood.'

Holroyd's gun coughed two bullets, and Jacobs fell sideways.

'Look on the bright side. At least you won't need to worry about a hangover.'

---

HOLROYD IS PLEASED with the progress on her to-do list.

Next: *Pendle.*

FORD OPEN PRISON: *A little, low-level outpost of relative freedoms. But ultimately, the inmates are still caged. Pendle, she saw through the wire fence, is tending to dry soil. Putting canvas over his patch, before excessive sun or floods try to destroy it. Nature isn't a friend, but then, humanity hadn't always been a friend to nature.*

She recalled Pendle saying in his early briefings, 'The prison population is responsible for a huge drain on our economy. And criminals get to share their skills and experiences with others there. It's like a finishing school for criminals, with many going on to re-offend. The courts: they're flooded with cases, and the stripped-down police force has escalating violence from criminals to contend with, leading to more hospitalisation or even death.'

He'd added. 'It's a war the police are losing. That's why the NSSA has been given the power to dispose of them. You'll be doing an excellent service by ridding the Nation of these people. Welcome to *Project Deadhead.*'

HOLROYD THOUGHT that *he who lives by the sword should not be surprised when it came to dying by it, too.*

*And it's personal: he'd been the root cause of Grey's death.*

*Even when love had stared her in the face, she could not recognise it. She did not want to expose her buried vulnerability to it. If she admitted and felt it, then everything she had done would fall into focus; she would see herself as the opposite of love, and wouldn't that make her evil?*

She drew that strand of thought back inside her.

Holroyd focused tightly on Pendle through the lens of her rifle.

Squeezed the trigger.

Pendle looked down, surprised at the concentric circles of blood growing from his chest. *He had the audacity to look angry, as if no version of the truth would allow him to deserve this: Pendle even died with arrogance.*

---

HOLROYD EASED into this prone position daily, waiting for the rabbit - the newly elected PM Ella Peterson - to appear from the emergency bunker in deepest Somerset.

Holroyd's scars itched, her injuries still knitting, healing, and the passing of the ache-inducing rainy season had helped. However, the lack of movement in her viewing point caused her discomfort.

Holroyd watched a cat move in the undergrowth, luxuriating in the early morning gentler sunshine. Rare dew droplets refracted bright light on the short brown grass that drank eagerly at the water. The cat's whiskers twitched; its eyes focused on the robin hopping about its positive, cheerful business. The sun must have made the bird lazy and off guard, because the cat caught it easily. Full head-on. So now the wings splayed out, its little twig legs and feet splayed out, everything visible except its head, which is firmly wedged within the cat's mouth. The bird's wings spasmed - *it is still alive.*

The cat is walking in Holroyd's direction, sees her and raises its tail in pleasure. It's looking pleased with its morning work. The cat probably did not think of itself as bad; it merely followed a natural DNA path laid out for it. *Holroyd could identify with the cat – her natural inclination towards this work made her job of killing so much easier.*

And Holroyd had to be sure; if she got a chance, she had to get it right like the cat. *Because, if she failed, Peterson's security would be cranked up, making getting at Peterson more problematic. Assuming she got away.*

*So far, no opportunity had offered itself, yet there was a chink in Peterson's armour.*

*Peterson insisted she runs, taking the risk of going outside, around the bunker grounds. But she is not wholly reckless; she picks random times each day. So, for Holroyd, it had been weeks of lying outside in surveillance, choosing a different set of hours each day, sometimes in extreme heat, in the hope of catching her target exposed.*

And this is Holroyd's lucky day.

Peterson emerged in a peaked cap with a circled target on her back, a humour Holroyd could appreciate. The PM took off along the path that hugged the barbed-wired fences.

Holroyd flexed her fingers, circulating blood. *All the better to feel the trigger.*

Every day, Holroyd had done nothing but prepare for this one shot. She constantly accounts for wind, visibility, temperature and gauges her state of self and how that would affect her pull on the trigger's metal curve.

Peterson's back is a plain target. No heat haze to obscure. Clarity of sight is perfect. She applied pressure. She eased her breath. Her body rolling with the recoil of the rifle.

The woman's body flew. Arms flailing behind her as if pulled back on wires, going down hard. White track shoes upended. Through the magnifying rifle lens, Holroyd saw the indented soles of the shoes where Peterson had settled. Unmoving. She reflexively wiped the lens.

She kept watching the downed figure. *Needing to be sure Peterson was dead.* But it's expensive to wait, which gives the security services a greater chance to catch her. The line of sight of the shot and the direction of the sound had given Peterson's team clues about where Holroyd was holed up. Because within seconds, cold-eyed, angry operatives surrounded her. Even for them, this was way too quick.

*Did they know what she was doing? If so, why didn't they stop her before she fired at Peterson?*

*The good news is they didn't have a kill order because she's still breathing.*

It didn't matter to Holroyd. She had achieved what she had set out to do, to kill one of the most dangerous, sociopathic people she had ever encountered—the *last one on her list.*

She dropped her rifle, receiving fist blows to her face as her reward.

'Gentleman, is the rough stuff necessary to assert your masculinity?'

She's brusquely searched. 'Not even a *please*?' she said. Her comment earned her more punches and a split lip. 'How about a one-on-one?' she offered.

Her aggressor looks prepared to dole out more retribution but is told to stand down by the group leader.

'Enough. We need to leave something of her for the interview.' His words, pregnant with contempt.

Then she's taken, blood trickling, her skin busy generating purple bruises, into the bunker. 'It's cool here. That's nice,' she said.

The hallway is narrow, low-ceilinged, and sharply lit, giving the illusion of getting smaller the deeper they enter the building. *Not a place for the claustrophobic.*

WHAT WERE *they going to do to her for killing the PM? Hang her? Shoot her? Maybe a spot of interrogation and torture? Another fact that would harm the sensitivities of the liberal classes, that the British would sink to such depths. But then, what the public didn't know wouldn't hurt them.*

She performed well under army interrogation training, often outlasting her peer group. *But everyone ultimately broke. It's only a matter of how long it took. Not that you were told that at the start of the training. She still looked forward to lasting as long as possible–it's always good to aim for growth.*

What Holroyd isn't prepared for is being bundled into a minimally furnished room with a picture of the monarchy hung on the wall and a decent plush carpet on the floor.

*Or to see Peterson, alive and well, standing behind a mahogany desk.*

'I see you are suitably shocked to see me,' said Peterson. 'How gratifying. Please take a seat. I have long been waiting for you to appear. Predictably, you would attempt to kill me in some misguided sense of loyalty to your country - and to your ex-partner, Grey.'

She mixed a little acid into Holroyd's wounds with her half-smile.

'Only my decoy has a sense of loyalty to her country, too. Fortunately for her, her bulletproof vest saved her life.'

'Fuck,' said Holroyd. 'Everyone does your dirty work for you, right? Who'll do the job of disposing of me?'

Petersen picked up a gun from her desk.

Holroyd held herself in readiness for the killing shot.

'Trust me, I'm tempted, but your punishment is to sit on death row. To wake each day, wondering if this is the day you die.'

Petersen smiled at Holroyd's dismay at her inhuman punishment.

'Others will try to destroy me. I'll take great pleasure in beating them.

Because I am strong enough to do whatever it takes.

To hold on to power.

To cut the dead away.

After all, there's an art to deadheading.'

# Acknowledgments

The Writer's HQ (Brighton), whose potty-mouthed, positive cajoling and flash fiction exercises taught me what it felt like ***to finish something***. And for the boundlessly generous and caring writers who gave their feedback and encouragement.

Louise Dean at The Novelry for her courses and the outstanding, patient tutors; Tasha Suri, Kate Riordan, Mahsuda Snaith and Amanda Reynolds, who led me to become a better writer, especially teaching me ***not to get in the way of the story***.

Lee MacKenzie (Neon Books), Kathy Hoyle (WHQ) and Craig Leyenaar (The Novelry) for their incredible editing skills and insightful feedback on my work.

Carla Jenkins for her Raw Writing group in Exeter, giving me contact with talented people such as Jem Sugden (for his exploratory design work).

Mark Swan of Kidethic who is responsible for creating this book's stunning cover design and the company logo.

The South-West Writers Group (Facebook) for being such a source of knowledge and willingly shared experience.

Fennec Adams (MBACP) for helping me bring DCI MacGillivray's psychological profile to life.

Ricky Ross for his unwavering support, who always assumed it was a case of *when* rather than *if* I would finish this.

And a big thanks to my family and friends for their support, for there are too many to name.

# About the Author

Mentor and coach, trail runner, hiker, skier, French language mangler, book reader, junior football league volunteer, Brighton and Hove Albion FC supporter, player of the occasional game of table football and a student of writing.